Black Holes in Space

Books by Patrick Moore

The New Guide to the Planets

A Survey of the Moon

Naked-Eye Astronomy

The Amateur Astronomer's Glossary

Amateur Astronomy

The Sun

Suns, Myths, and Men

How to Make and Use a Telescope, by Patrick Moore
  and H. P. Wilkins

Life on Mars, by Patrick Moore and Francis Jackson

Craters of the Moon, by Patrick Moore and
  Peter Cattermole

Can You Speak Venusian?

The Yearbook of Astronomy (annual)

Astronomical Telescopes and Observations for
  Amateurs

Black Holes in Space

Patrick Moore and Iain Nicolson

W · W · NORTON & COMPANY · INC ·

NEW YORK

First American edition 1976
Library of Congress Cataloging in Publication Data
Moore, Patrick.
    Black holes in space.
    Includes index.
    1. Black holes (Astronomy). I. Nicolson, Iain, joint
author. II. Title.
QB843.B55M66   1976   523   75-15604
ISBN 0-393-06405-0

Printed in the United States of America
  ´ 2 3 4 5 6 7 8 9 0

Contents

Black Holes in Space

1 The Black Hole

There are more things in Heaven and Earth, Horatio . . .

This quotation is hackneyed enough, and yet it is more true today than it has ever been. Not long ago we were confident that even though we did not know all the secrets of the universe, at least we knew all the different types of objects to be found in the sky. There were stars and nebulae, star-systems or galaxies, planets and satellites; everything seemed reasonably clear-cut. Now, in 1974, the whole situation has altered. We have strong evidence of objects so weird it is hard to form any picture of them. And of these the most fantastic of all are Black Holes.

A Black Hole is a region of space (*not* a solid body) into which matter has fallen, and from which *nothing*, whether material objects or even light itself, can escape. If Black Holes exist, we can never see them directly since they neither emit nor reflect light. They truly live up to their name!

Yet it may be possible to detect them by their effects on other bodies; for although they can emit no signal they still have gravitational fields. Even as we are writing, strenuous efforts are being made to identify Black Holes in the universe. There is the strange object in the sky, for example, known as Cygnus X-1, which is sending out tremendous quantities of X-rays into space. This object has been identified with a powerful blue star, accompanied by an *invisible* companion which itself seems to be ten or twenty times more massive than our Sun. It is very difficult to account for this invisible object except by invoking the idea of a Black Hole. The X-rays give another clue: they are just what would be expected if matter was being drawn towards a Black Hole before being sucked in and lost for ever.

In the vicinity of a Black Hole, and even more so inside one, conditions become so strange that to describe them in everyday language is wellnigh impossible. Our common sense notions and our cherished scientific laws take a very heavy beating, and right in the centre of a Black Hole they cease to have any meaning at all. We usually think of time as passing at a uniform rate, but where Black Holes are concerned we have to deal with the most paradoxical situations. As we shall see, if you were to fall into a Black Hole, you would quickly seem to reach the centre of it; yet someone who wisely stayed outside would be quite convinced that you never reached the centre at all. Who is 'right'? Or again, what happens to matter inside a Black Hole? Theory suggests it is *crushed out of existence*. These notions are an affront to common sense, but in the following chapters we hope to show that this is indeed the way the universe operates.

Do Black Holes really exist? If so, how were they created? How can they be detected? What would happen if we encountered one? These are some of the questions we shall try to answer as best we can. However, let us say right at the outset that there is as yet no *conclusive* evidence that Black Holes exist. On the other hand it would be a most puzzling thing indeed if it were ever shown that Black Holes do not exist somewhere in the universe. Strange as it may seem, to deny the possibility of Black Holes would require a major upheaval in scientific theory.

Objects which appear to contradict common sense turn up in other branches of science apart from astronomy. For example, in the 1930s a 'particle' called the *neutrino* was proposed by nuclear physicists. It had no mass and no electric charge, and could not be directly detected. The idea of a particle with no mass hardly fits in with everyday ideas, yet its effects on other particles could be predicted. These effects were indeed observed, and it was concluded that the neutrino did indeed exist. (In fact, nowadays, astronomers can measure neutrinos coming from the very core of the Sun, and use these to check the temperature there.)

If Black Holes do exist, the most likely way for them to originate is from the death throes of the most massive stars in the universe. Weird though they seem, they may be the natural by-products of the way stars evolve, and there may be a lot of them around. They should be regarded, therefore, in the context of the universe as a whole, and because of this we approach the subject with a description of the universe at large and the types of bodies that exist within it. This is followed by the life-story of a star as we understand it at present, and an outline of 'Pulsars' – those remarkable sources of radio pulses which now seem to be identified with neutron stars, incredibly dense bodies which themselves represent the final fate of certain kinds of stars.

Only then do we launch into what lies behind the Black Hole concept, how they may be formed, and ways in which they might be detected. Finally, we branch out to look at some of the wider aspects of Black Holes. Can they explain some of the most mysterious and most powerful sources of radiation in the universe? Can mini-Black Holes exist? Is the universe itself a Black Hole? The possibilities and speculations are limitless.

If the reader feels that he is well into the book before really meeting up with Black Holes as such, we can only say that they cannot be considered in isolation. If they exist they are as much a feature of the universe as the familiar stars and planets. Whether or not any matter that falls into Black Holes is still a part of our universe is quite another question.

The narrative, we hope, will give an idea of the Black Hole story as it stands today. Much of it perhaps sounds like science fiction, but nowadays science fiction has a habit of turning inexorably into science fact. The Black Hole concept demonstrates once again the validity of the statement by J. B. S. Haldane: 'The universe is not only queerer than we imagine, it is queerer than we *can* imagine.'

2   The Universe Around Us

Before starting our investigation into Black Holes or any other exotic objects, it is essential to set the scene, and to give a general account of the universe in which we live. Compressing any such account into a few pages is bound to lead to omissions and over simplification, but it will be better than nothing. So let us begin near home, and consider the Sun's family or Solar System.

The Sun is an ordinary star. It is a globe 865,000 miles in diameter, which may also be expressed as 1,400,000 kilometres (for those who prefer the metric system into which everyone is being dragooned). It shines by its own energy; there is nothing the least remarkable about it, and it seems so glorious in our skies only because it is so close to us. The distance between the Earth and the Sun is 93,000,000 miles (or around 150,000,000 kilometres); this is also expressed as one astronomical unit. It may sound a long way, but it is cosmically insignificant. Even the nearest star beyond the Sun lies at a distance of over 40 million million kilometres.

Moving around the Sun we have the nine planets: Mercury and Venus, closer to the Sun than we are; then the Earth, and then Mars, last of the inner group. Beyond we come to the four giants (Jupiter, Saturn,

Not-Planet

Uranus and Neptune) and then the rather enigmatical Pluto, which is almost certainly smaller than the Earth and which seems to be in a class of its own insofar as planets are concerned. Unlike the stars, the planets have no light of their own, and shine only because they are being lit up by the Sun. In the admittedly unlikely event of the Sun being suddenly snuffed out, the planets too would cease to shine – and the same would apply to the Moon and the various satellites in the families of other planets.

Practical space research has sent men to the Moon and automatic probes out to the inner planets and Jupiter. We can now study the huge Martian volcanoes as they have been photographed from close range; in December 1973 Pioneer 10 by-passed mighty Jupiter, sending back information which we could never have hoped to obtain so long as we contented ourselves with making observations from the surface of our own world. Eventually, provided that *homo sapiens* does not destroy civilization by indulging in nuclear warfare, there is no reason to doubt that astronauts will penetrate to the very boundaries of the Solar System. That, however, is as much as we can say at present. Rockets of the Apollo, Pioneer or Mariner type will never take us to the stars, unless we are prepared to consider a journey lasting for many thousands of years (which means that anything of the kind is so wildly speculative that it is not worth following up as yet). And because the other planets in the Solar System do not seem to be suitable for advanced life, we must resign ourselves to the fact that there are no immediate prospects of contacting the many other intelligences which must certainly exist in the universe. The distances are too great. If contact is to be made, it must be by some method about which our present ignorance is complete; and yet we may even have to wait for the mountain to come to Mahomet instead of *vice versa*. Time will tell.

Though there is still a great deal of discussion as to how the Earth and the other planets came into existence, we may be fairly sure that the Solar System originated between 4,500 and 5,000 million years ago. The age of the Earth has been measured by several reliable methods, all of which lead to the same conclusion, and analysis of the rocks brought back from the Moon is also in agreement. It also seems highly probable that the planets (and satellites such as the Moon) were built up by accretion from particles once making up a 'cloud' associated with the Sun. What can happen to the Sun can also happen to other stars, and it is logical to assume that planetary systems are commonplace. We have even some positive evidence of them. Every cosmical body pulls upon every other cosmical body; for instance, the motion of the Earth round the Sun is controlled by the Sun's very powerful pull of gravity, but is also affected to some degree by the much weaker pull of Venus, Mars and the other planets. With a comparatively lightweight star attended by a very massive planet, it is possible to detect the tiny 'wobblings' of the star due to the pull of its invisible attendant. Wobblings of this kind have indeed been found for several relatively nearby stars. Unfortunately we cannot hope to see the planets of other suns, because they are too small and too faint.

The concept of an invisible body pulling upon a visible one, and dragging it out of its predicted position, is of tremendous importance in any discussion of Black Holes, which is why we have introduced it here. But to delve further into the numbers of planetary systems, or the chances of life upon them, would be too much of a digression; so let us leave the Solar System, and look at the star system or Galaxy.

At once we have to deal with immense distances, and the conventional mile or kilometre is inconveniently short. Luckily, Nature has provided us with a better unit. Light moves at 186,000 miles or 300,000 kilometres per second; therefore in a year it covers rather less than 10 million million kilometres, and this is our agreed unit, known as the *light-year*. The nearest star beyond the Sun (Proxima Centauri) is $4\frac{1}{4}$ light-years away; Rigel in Orion is about 900 light-years away, and so on. Our knowledge of the universe beyond the Solar System is always bound to be out of date. When we look at Rigel, we see it not as it is now, but as it used to be 900 years ago. Inside the Solar System, of course, the light-travel time is much less; we see the Sun as it used to be a mere $8\frac{1}{2}$ minutes ago.

Our Galaxy contains around 100,000 million stars, each of which is a sun in its own right, and many of which are a great deal larger, hotter and more powerful than our Sun. Thousands of years ago, the ancients divided up the stars into groups or constellations, and gave them names; who has not heard of the Great Bear, Orion, and the celestial Scorpion, for instance? It was known that the constellation patterns do not change noticeably even over periods of many lifetimes, and we still sometimes meet with the old expression 'Fixed Stars'. Only the Sun, Moon and planets wander around the sky from one constellation into another, and even these keep to a well-defined belt in the sky called the Zodiac.

In fact, the stars are anything but fixed. They are moving around in space at high speeds; if they seem to be motionless, as seen with the naked eye, it is only because they are so far away from us. With modern equipment the individual or 'proper' motions of the stars can be measured, but the shifts are very slight indeed, and to all intents and purposes the patterns we see today are the same as those which must have been seen by Julius Cæsar or Homer. The apparent east-to-west motion of the sky is due entirely to the real rotation of the Earth upon its axis, and has nothing to do with the stars themselves.

The Earth, as we have seen, is over 4,500 million years old, and the Sun is presumably older. If the Sun were burning, it could not go on giving out energy for anything like this period; and in any case it is too hot to burn. Even the surface temperature amounts to 6,000 degrees Centigrade, and near the core the temperature rises to the amazing value of some 14,000,000 degrees. We must look for another source of energy, and we find it in nuclear transformations. This problem will be discussed in Chapter 3; for now, it is enough to say that the Sun is changing one element (hydrogen) into another element (helium), giving off energy and losing mass as it does so. Each second of time, the Sun loses 4,000,000 tons of mass. We are delighted to be able to assure you, however, that there is no cause for panic on this score; the Sun will not change much for five or six thousand million years in the future.

The Sun, let us repeat, is a normal star, and astronomers are ungallant enough to classify it as a Yellow Dwarf. Some of the other stars in the Galaxy are much more powerful, and are white-hot; yet others are dimmer and redder. There is plenty of variety in the grand population of a hundred thousand million. The Galaxy also contains clusters of stars; for instance, many people will know the Pleiades or Seven Sisters in the constellation of Taurus (the Bull), in line with Orion's Belt. At least seven individual stars are visible with the naked eye on a clear night, making up a beautifully compact group, and telescopes raise the total to over two hundred. There is little doubt that the stars in an open or loose cluster of this sort have a common origin, and came into existence at about the same time.

Quite different are the gaseous nebulæ, of which the finest example is in the Sword of Orion, just below the Belt; it is known as M.42, because it was the 42nd object in a famous catalogue drawn up in the year 1781 by the French astronomer Charles Messier. Look at it with the naked eye, and it appears as a misty haze. Telescopically it gives the impression of a mass of swirling gas, and appearances are not deceptive. M.42 is genuinely gaseous, though the material in it is amazingly rarefied (many millions of times less dense than the air we breathe). It is thought that fresh stars are being born out of the material in M.42 and similar nebulæ, so that we are looking at true stellar birthplaces.

Next, what about the Milky Way, that glorious band of light which stretches across the sky from horizon to horizon? There are many old legends about it, but not until the invention of the telescope was it found to be made up of stars. This was discovered by Galileo in the winter of 1609–10, when he turned his primitive 'optick tube' to the sky. Any modern binoculars will show the same aspect; it looks almost as though the stars are close enough to bump together.

This is not so. The stars are very widely separated in space, and the chances of a head-on collision are so slight that we can ignore them (there would be much more danger of a collision between two gnats flying around inside a large concert hall). We now know that the Galaxy is a flattened system, as shown in the diagram; it has a maximum diameter of about 100,000 light-years, and the Sun, with the Earth, lies not far from the central plane, at about 33,000 light-years from the nucleus. Look along the main plane of the Galaxy, towards A or B, and you will see many stars almost in the same line of sight; this accounts for the Milky Way appearance. We cannot see all the way through to B, beyond the actual centre, because of the obscuring material in the way. In fact, we cannot even see as far as the nucleus itself. There have been suggestions, admittedly speculative, that there may be a Black Hole in those hidden regions. . . . The direction of the galactic centre lies toward the magnificent star-clouds in the constellation of Sagittarius, the Archer, visible low down in the south from England during summer evenings.

If the galactic nucleus is permanently hidden, as though by a sort of cosmical fog, it might be thought that we could never learn anything about it. Fortunately this is not so. Light-waves can be, and are, blocked; but they represent only a small part of what we call the electromagnetic spectrum.

Light may be regarded as a wave-motion, and its colour depends upon its wavelength. Thus the longest-wavelength light that we can see is red, and the shortest is violet. The wavelengths are very short indeed – tiny fractions of a millimetre – and are measured in units known as Ångströms (Å) in honour of their discoverer.*
One Ångström is equal to a ten-millionth of a

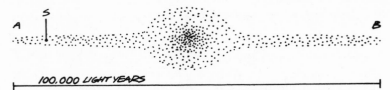

Plan of the Galaxy
S represents the Sun. Looking along the main plane of the system (toward A or B) produces the Milky Way effect.

*Anders Ångström, the 19th-century Swedish physicist. It was inconvenient of him to have a name beginning with the purely Swedish letter Å!

The Electromagnetic Spectrum

Visible light takes up only a very small part of the whole range, from long radio waves to very short gamma-rays. Only radiations in the 'radio window' and the visible range can come to us direct from space; all the rest are blocked by the atmosphere, so that instruments must be taken up in rockets or artificial satellites if these radiations are to be studied.

millimetre, and the range of visible light is from about 7,500 Å (red) down to 3,800 Å (violet). Radiations shorter than this cannot be seen, because they do not affect our eyes; we have the ultra-violet, then the X-rays and then the remarkably short gamma-rays. On the longer-wavelength side of the visible band we have the infra-red and then radio radiation, where the wavelengths may amount to many metres. The diagram shows that the visible range is extremely small, and so long as astronomers had to depend entirely upon it they were badly handicapped.

Fortunately, objects in the sky send out radiations at all wavelengths – not only in the visible range. In the early 1930s an American radio engineer named Karl Jansky discovered that with his improvised aerial he was picking up radio waves from the Milky Way, and by now radio astronomy has become a vital research tool. The flux collectors are called 'radio telescopes', and are of various patterns, some of which look decidedly bizarre. With the familiar dish type, such as that at Jodrell Bank in Cheshire, the long-wavelength emissions are collected by a huge metal bowl, in this case 250 feet (76 metres) in diameter, and are brought to focus. Naturally, no visible picture is produced, and one cannot look through a radio telescope; but in its way, a trace on a graph can be just as informative. Without radio astronomy, our knowledge of the universe would still be restricted.

Now let us go back to our flattened Galaxy. Light-waves are blocked out by the interstellar material. Radio waves are not; and we can therefore obtain signals from regions which we can never see directly. There are also various discrete sources of radio waves in the Galaxy – mainly the results of past stellar explosions – about which more will be said below.

When Charles Messier drew up his catalogue of over a

hundr.ed star-clusters and nebulæ, he included objects of various kinds. There were the open clusters, such as the Pleiades; there were also the much more compact, symmetrical objects known as globular clusters, of which the brightest example visible from Britain is M.13 in the constellation of Hercules; and there were gaseous nebulæ, as well as other nebulæ which looked as though they were starry in nature. More than fifty years ago Harlow Shapley, the American astronomer, who will always be remembered as the great pioneer of Milky Way studies, measured the distances of the globular clusters and found that their distribution was lop-sided as seen from Earth; this led him on to a good estimate of the size and shape of the Galaxy. But the starry nebulæ were different altogether, and nobody could be sure of their nature. Originally, Shapley believed them to be members of our own Galaxy; but they were so remote that all ordinary methods of distance-finding were useless.

Some of them, notably Messier's 31st object (in the constellation of Andromeda), were known to be spiral in form; this had been discovered as long ago as 1845 by an Irish astronomer, the third Earl of Rosse, who had built a huge reflector with a 72-inch mirror and had used it to make some amazing discoveries.* They were not like the gaseous nebulæ, but positive proof had to wait until 1923, with some pioneer work by Edwin Hubble at the Mount Wilson Observatory in California. Hubble found that the starry nebulæ contained stars of a very special

---

*This episode is one of the most remarkable in the history of science, and remains unique. Alone and unaided, Lord Rosse built what was then much the largest telescope the world had ever seen. For the full story see *The Astronomers of Birr Castle*, by Patrick Moore (Mitchell Beazley, London 1971).

kind, known as Cepheid variables. Unlike most stars, these Cepheids do not shine steadily for year after year; they brighten and fade over short periods. Delta Cephei, the star after which the class has been named, has a variation period of 5.3 days; this is the interval between one maximum and the next. The variations are very marked, though the star is always bright enough to be visible with the naked eye.

For reasons which were then unknown (though the situation is rather clearer now), the real luminosity of a Cepheid depends upon the length of its variation period; the longer the period, the more powerful the star. In fact, a Cepheid *tells you its real luminosity by the periods of its fluctuations in light*. This means that Cepheids can be used as 'standard candles' for measuring distance by comparing their real with their apparent luminosity. Hubble found that the Cepheids in the Andromeda Spiral were so remote that they could not possibly lie inside our Galaxy. We now know that the Spiral is over two million light-years from us, so that we are seeing it as it used to be more than two million years ago.

This work of Hubble's was probably the most important breakthrough since the invention of the telescope. It showed that not only is the Sun unimportant in the Galaxy, but that the Galaxy itself is unimportant in the universe! The Andromeda Spiral is a considerably larger system, containing more than our own quota of a hundred thousand million stars.

It followed, of course, that the other 'starry nebulæ' must also be external systems, and this name was therefore dropped, to be replaced by the much more appropriate 'galaxy'. The Andromeda Spiral is one of the very nearest, and is a member of what we call the Local Group, of which other members are the rather smaller spiral in Triangulum and the two naked-eye southern systems

called the Magellanic Clouds – invisible, alas, from anywhere in Europe.

Much later, it was found that the clouds of cold
rarefied hydrogen spread through the Galaxy send out
long-wavelength radiations of 21 centimetres wavelength.
Radio astronomers were able to plot the distribution of
the hydrogen, and it has been confirmed that our
Galaxy too is a spiral, so that if seen from 'above' (or
'below') it would look like a rather loose Catherine-
wheel. There are spirals of all sorts in the sky; some
tightly-wound, some very open; some with curious bars
across them; some (the Seyfert galaxies) with very bright,
condensed nuclei and faint arms. It would be idle to
claim that we have as yet any real idea why spiral arms
form, or how they evolve. Yet not all the outer galaxies
are spiral; some are elliptical, some spherical, and some,
such as the Magellanic Clouds, irregular in form.

Cepheids are powerful stars, and can be seen out to
distances of millions of light-years, but with the more
remote galaxies they fade into the general background
blur; eventually we cannot even make out the powerful supergiant stars, which also can be used as standard
candles (though they are much less uniform and much
less reliable). To penetrate still further in our 'space
soundings', we have recourse to instruments based upon
the principle of the spectroscope.

A telescope collects light; a spectroscope splits it up.
Newton, as long ago as 1666, found that when he passed
a beam of sunlight through a glass prism, the result was a
rainbow band or spectrum, with red at one end and
violet at the other (as show in the diagram). In fact, the
beam of sunlight is not 'pure'; it is a mixture of all colours – and these are bent or refracted unequally by the
prism. Red is bent less than orange, orange less than
yellow, and so on through green, blue, indigo and violet.

Newton never took matters much further, but in the early 19th century a German optician, Josef Fraunhofer, studied the spectrum of the Sun, and found that the rainbow band was crossed by dark lines. Later still, Gustav Kirchhoff, at Heidelberg, found the reason for the lines. An incandescent solid, liquid or gas at high pressure will yield a rainbow or continuous spectrum; a gas at low pressure will produce an 'emission' spectrum made up of isolated bright lines. Each line is due to some particular element or group of elements, and cannot be duplicated by any other substance. With the Sun, we have a continuous spectrum from the bright surface or photosphere; above this is a layer of much thinner gas, which produces a line spectrum – but because of the presence of the background, the lines appear dark instead of bright. Their positions are unaltered, and so they can be identified. For example, two prominent dark lines in the yellow part of the rainbow correspond to two bright yellow lines due to sodium, observable in the laboratory, so that we can tell that there is sodium in the Sun.

Quite apart from this, the spectrum can give invaluable information about the motion of the light-source, because of the familiar Doppler effect. If a light-source is approaching, more light-waves per second will reach you than would be the case if the source were standing still; the wavelength seems to be shortened, and the object looks 'too blue.' If the light-source is receding, the wavelength appears to be lengthened, and the source looks 'too red'. The actual colour-change is much too slight to be noticed in our everyday experience, though the well-known rise and fall of the whistle of a train as it passes you is due to just the same principle. However, when we turn to the spectra of astronomical objects, the Doppler effect shows up as a shift in the dark lines; for a body which is moving toward us, all the dark lines will be shifted toward the violet or short-wave end of the spectrum; with a receding body, there will be a Red Shift. We give this capital letters because Red Shifts are so vitally important in all modern research into the nature of the universe!

Apart from the few members of our Local Group, all the galaxies show Red Shifts in their spectra; and this means that all of them are racing away from us. Moreover, the greater the Red Shift, the greater the speed of recession, and it has been found that the rule is 'the further, the faster'. Every group of galaxies is receding from every other group, and the whole universe is expanding. The most remote galaxies known to us are at least 5,000 million light-years away, and are moving outward from us at appreciable fractions of the velocity of light.

This seems incredible, and there are a few astronomers who doubt whether it is true. Remember, we depend almost entirely upon the Red Shifts in the spectra of galaxies, interpreted as Doppler effects. If the Red Shifts are not Dopplers, but have some other explanation, then all our measurements of remote galaxies may be wrong, forcing us to do some very serious re-thinking. This, however, is very much of a minority view, and according to the overwhelming majority of modern astronomers the Red Shifts *are* Dopplers; the galaxies really *are* receding at these fantastic rates; and the universe *is* expanding. But it would be premature to claim that the subject is definitely closed, particularly in view of some recent, rather disturbing discoveries.

There are some galaxies which are extremely powerful emitters of long-wavelength radiations, and which are therefore known as radio galaxies. Most of these seem to be either of the Seyfert type (bright nuclei, faint arms) or to have suffered great internal disturbances. After the end of the Second World War, when radio astronomy really came into prominence, energetic efforts were made to catalogue all the detectable individual sources, and to identify them with objects which could be seen in ordinary telescopes. Some of the sources were found to lie inside our Galaxy – notably the famous Crab Nebula, to be described below. Others lay beyond our own system, and were linked with radio galaxies such as the almost spherical M.87 in Virgo, which has a strange 'jet' issuing from it. Yet there were also some sources which corresponded in position with what looked like faint blue stars. This was certainly very odd, particularly as these 'stars' displayed unusual spectra with lines which defied identification.

A radio telescope cannot pinpoint the position of a radio source as accurately as an ordinary telescope can give the position of a visual source, and for some time the situation remained uncertain. Luckily one of the puzzling sources lay in a position where it could be hidden or occulted by the Moon. The Moon's position is known very precisely for all times; therefore, the moment of the cutting-off of the waves from the radio source as the Moon passed over it would lead to a determination of the actual location of the source itself. At Palomar, Maarten Schmidt was provided with this information, from observations made with the large radio telescope in Australia. In 1963 he made a new check of the spectrum of the 'star' which lay in the right position, and found, to his surprise, that the object was not a star at all. The spectral lines were shifted so far to the red that the object itself presumably lay at a tremendous distance – thousands of millions of light-years. Yet it looked small and vitually stellar. This was the first identified 'quasar'; by now hundreds have been catalogued.

Quasars are undoubtedly of immense significance,

but we cannot pretend that we really know what they are. If the Red Shifts in their spectra are ordinary Doppler effects, then some of them are even further away than the remotest known galaxies – we must reckon in the distances of well over 5,000 million light-years in some cases, with velocities of recession exceeding 90 per cent of the velocity of light. It follows that a powerful quasar must radiate as strongly as, say, 200 whole galaxies put together – that is to say, of the order of 200 × 100,000,000,000 Suns. It seems absurd; how can so much energy come from a body only a light-year or so in diameter at most? All sorts of explanations have been offered, ranging from chains of tremendous stellar explosions to gravitational collapse, or even the mutual annihilation of ordinary matter and what is termed 'anti-matter'. But as yet the quasar riddle remains unsolved.

As might be expected, there is a measure of disagreement about the distances of quasars. In America, Halton Arp has studied their distribution, and has found that in some cases they are lined up with systems of galaxies, even though they show different Red Shifts. What may be even more significant is a recent (1973) discovery of a pair of quasars which lie side by side, seeming to be genuine twins, and yet have dissimilar Red Shifts. If the Red Shift measurements are unreliable for giving us the distances of quasars, then they may also be unreliable for remote galaxies. Certainly this new research seems to have made the whole situation even more baffling than before, and we can only wait and see what happens next. In any case, quasars are extraordinary objects, and they must lie well beyond our Galaxy even if they are not at what are often termed 'cosmological' distances; and their sources of power remain as enigmatical as ever. We are not even sure whether they are the nuclei of galaxies of very special type. And with quasars, too, it has been suggested that there may be central Black Holes.

Much has been heard of late about 'the origin of the universe', but the term is, in a way, rather misleading. Nobody has ever been able to suggest how the material making up the Earth, the Sun, the stars and the galaxies came into existence. All we can do is to start from some basic assumption, and then work out a sequence of events ending up with the universe as we see it today. Either the material came into being at some particular moment in time, at least 10,000 million years ago, or else it has always existed, in which case the universe has an infinite past. Nowadays there are two basic theories, the 'evolutionary' and the 'cyclic'.

According to the evolutionary idea, all the matter of the universe was suddenly created in the form of what may be called a 'primæval atom'. This atom exploded, sending material outward in all directions. Galaxies were built up; inside the galaxies, stars formed; and with the stars, planets. The universe had a definite beginning, and is evolving toward eventual death. Expansion will never stop.

But do the speeds of recession go on increasing with distance? The Red Shifts can be used to measure the velocities at which the galaxies are racing away, and there is a definite relationship, known (appropriately) as Hubble's Law. Yet it may be that as the recessional velocities approach that of light, the rate of expansion slows down. If so, expansion may eventually stop, after which the galaxies will start to come together again. Finally, in about 60,000 million years (the exact period is bound to be somewhat uncertain!) all the galaxies will meet in catastrophic collision, producing another 'Big Bang', after which expansion will begin anew and the whole story will be repeated. On this cyclic pattern, the universe will never end; it will be periodically reborn.

Various other theories have been put forward. For some years the 'steady-state' idea was popular; on this hypothesis the universe has always existed, and will exist for ever, so that as old stars and galaxies die they are replaced by new systems, formed from material which is spontaneously created out of nothingness in some unexplained manner. The steady-state theory is philosophically satisfying, but unfortunately it has not stood up to observational tests. If it were true, then the galaxies would always have had the same sort of average distribution, and if we could have looked at the universe thousands of millions of years ago it would have seemed superficially very like the universe we see today. In effect, we can make this kind of observation, because the light and radio waves coming from objects thousands of millions of light-years from us were sent out thousands of millions of years ago – and to all intents and purposes we are looking into the past. Radio research is particularly valuable here, because it can detect more distant sources. It has been found that the distribution is not the same in the very remote regions as it is in our own part of the universe; therefore the steady-state picture is not valid, and almost all astronomers have now given it up, albeit with considerable reluctance.

All in all, the cosmological picture seems rather less straightforward now than it did before the discoveries of unexpected objects, notably the quasars. Whether we will ever solve the problem of what is called – for want of a better term – the 'Creation' remains to be seen. There is a great deal that we do not know or even suspect, particularly with regard to fundamentals. What is the precise nature of gravity? How far does 'space' extend – or is it infinite? And if infinite, can we visualize something which goes on and on and on for ever?

Let us repeat that this review of the observable universe is very incomplete, and various important topics (such as the background radiation which has been interpreted as the result of the original 'big bang') have not even been mentioned. But it will, we hope, serve as an introduction to our main theme. The next step is to say something about the life-histories of the stars, because if Black Holes really exist there is little doubt that they began their careers as very massive stars.

3 Red Giants, White Dwarfs and the Evolution of a Star

Look up into a brilliant night sky, and you may well think that all the stars look alike. Certainly they cannot be seen as anything more than points of light, even with the most powerful telescopes ever produced; they are too far away to show disks, despite their tremendous sizes. Superficially, a star is much less spectacular than a planet, which shows a great amount of detail. There are few sights which can match the beauty of the belts and satellites of Jupiter, the white polar caps of Mars or, perhaps above all, the rings of Saturn.

Yet a closer examination will show that the stars have characters of their own. In particular, they are of differing colours. Look at Orion, the brilliant constellation which dominates the evening sky during winter, and you will see that its two principal stars are clearly dissimilar; Rigel, at the lower right of the constellation pattern, is brilliantly white, while Betelgeux in the upper left is a lovely orange-red. The difference is due to surface temperature. Rigel is hotter than the Sun; Betelgeux has a surface below 3,500 degrees Centigrade, so that it is only red-hot, though in compensation it is much the larger of the two.

Telescopes alone can tell us only a limited amount about the stars, and most of our knowledge has been obtained by means of the spectroscope. The spectrum of a hot star is entirely different from that of a cool one, and during the last century attempts were made to classify the stars into various spectral types, each of which was given a letter of the alphabet. It would have been logical to start with the hottest stars and work through to

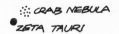

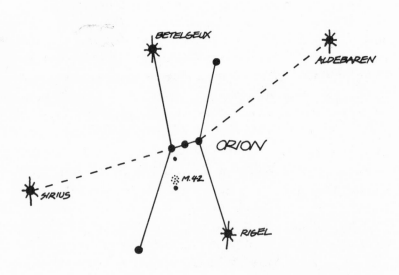

Orion
The two main stars are Betelgeux and Rigel; the Belt stars point one way to Aldebaran and the other way to Sirius.
The Orion Nebula is M.42, and the Crab Nebula, near Zeta Tauri, is M.1.

the coolest in the A, B, C . . . order, but inevitably the early efforts were faulty, and the final sequence was alphabetically chaotic. From hot to cool, the main types are now O, B, A, F, G, K and M. Of these, the O-stars are extremely hot, and bluish or greenish white; B are white; so are A; F and G are yellow, K orange and M orange-red. We also have stars of type W, which really lead off the sequence ahead of O, and types R, N and S, which are orange-red or red and which follow M.

It may be useful to look at a few typical stars. *Sirius* in Canis Major (the Great Dog) is of type A; it is a mere 8.6 light-years from us, so that it is one of our closest stellar neighbours; and it is 26 times as luminous as the Sun. Mainly because of its nearness, it shines as much the brightest star in the sky, and when low down it seems to flash all the colours of the rainbow – though twinkling or scintillation is due purely to the fact that the starlight has to come to us through our own unsteady atmosphere. Sirius has a white-hot surface, at over 10,000 degrees.

In Orion we have *Rigel* and *Betelgeux*, which have already been mentioned. Rigel, at 900 light-years from us, is at least 50,000 times as powerful as the Sun, and has a B-type spectrum; were it as close as Sirius, it would cast shadows. Betelgeux, of type M, is not so powerful as this, but it still outranks the Sun by a factor of several thousand, and its diameter is of the order of 250,000,000 miles (400,000,000 kilometres), so that it could swallow up the whole path of the Earth round the Sun. Yet its mass is only 15 times that of the Sun, because its outer layers are immensely rarefied, and correspond to what we would normally call a laboratory vacuum. All 'Red Giants' share this trait, and indeed the masses of the stars show a much lesser range than either the diameters or the luminosities. Usually, 'the smaller, the denser', so that we may draw an analogy of weighing a small lead pellet against a large meringue. (This applies to what we may call normal stars. Abnormal ones will be discussed below.)

Next, look at *61 Cygni*, the first star to have its distance measured. Here we have a 'Red Dwarf' – or, rather, a pair of Red Dwarfs. The two components, both of type K, are not only smaller than the Sun but are also much less luminous, so that they have nothing in common with the Red Giant star Betelgeux except their colour. They make up what is termed a binary system, with the two components moving round their common centre of gravity. Binary stars are extremely frequent in the Galaxy, and there are large numbers of pairs observable with a very small telescope.

In passing, note that a star's 'magnitude' is a measure of its apparent brilliancy; the lower the magnitude, the brighter the star. Thus Vega (magnitude o) is brighter than Deneb (1.3), and Deneb is in turn brighter than the Pole Star (2.0). People with normal eyesight can see down to magnitude 6; with the world's largest telescopes, photographs can be taken of stars down to magnitude 23. Obviously, apparent magnitude depends partly on the star's distance and partly on its real luminosity.

It used to be thought that a star shone by ordinary 'burning'. When this logical idea had to be given up (a very long time ago now) various other theories were proposed. One, which held the field for some time, involved gravitational contraction. It was supposed that a star began its career by condensing out of nebular material – that is to say, gas and 'dust' – and began to shrink; the interior would heat up, and the energy would make the newly-born star luminous. Unfortunately for this idea, the time-scale is wrong. We know that the Sun is at least 5,000 million years old, and many stars are presumably older. No simple contraction could

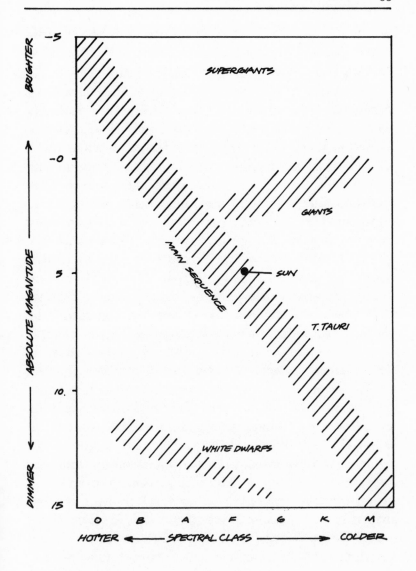

The H-R Diagram

keep a star radiating for so long, and another energy-source had to be found.

This is probably the right place to introduce what is known as the Hertzsprung-Russell or H-R Diagram, so named because it was due to the work of two astronomers: Ejnar Hertzsprung of Denmark, and Henry Norris Russell of the United States. Just before the first world war, Hertzsprung made a chart in which he plotted the stars according to their luminosities and their spectral types. In the H-R Diagram given here, luminosity is represented by 'absolute magnitude' – the apparent magnitude that a star would have if it could be seen from a standard distance of 32.6 light-years.* The Sun's absolute magnitude is 4.9; thus if observed from our standard distance it would be a very dim naked-eye object, whereas Rigel, with an absolute magnitude of minus 7, would far outshine any star in our sky.

It had already been found that red stars, and to a lesser extent orange and yellow ones, come in two varieties: very large and luminous, and very small and dim, so that there is a true division into Giants and Dwarfs. Even a casual glance at the H-R Diagram given here will bring this out. Most of the stars lie on a well-defined band extending from the upper left to the lower right; this is called the Main Sequence, and our Sun, with a spectral type of G and an absolute magnitude of around 5, is a typical Main Sequence star. The giant branch lies to the upper right. Look for red stars of type M which are about equal in luminosity to our Sun,

---

*In case anyone is wondering 'Why 32.6 light-years?' let us explain that this is equal to 10 parsecs; and 1 parsec is the distance at which a star would show a *par*allax of one *sec*ond of arc. It works out, of course, at 3.26 light-years, and actually no star except the Sun is as close as this.

and you will not find them; the red stars are either much brighter or much feebler.

When the first H-R diagrams were drawn, it was tempting to regard them as indicating an evolutionary sequence, and Russell put forward a theory which sounded delightfully plausible. A star would begin by condensing out of nebular material, and at first would be large, cool and red. It would contract, because of gravitation, and the inside would heat up. Eventually, the star would become hot enough for reactions to begin among the fundamental atomic particles such as protons and electrons. A proton, making up the nucleus of a hydrogen atom, has a unit charge of positive electricity; an electron, which orbits the nucleus, has a unit negative charge; so, according to Russell's theory, they could collide and wipe each other out, releasing energy. It was assumed that this annihilation of matter would suffice to keep a star shining for a very long period indeed, and it was calculated that the lifetime of a star could be as much as 10 million million years. It would become hot and white (type B) and then slide gently down the Main Sequence, becoming less and less energetic, until it reached the Red Dwarf stage (type M, lower right), after which it would lose the last of its heat and would become a cold, dead globe.

This sounded reasonable enough, but various difficulties soon began to make themselves unpleasantly obtrusive. For one thing, the time-scale was as clearly too long as the original one (gravitational contraction) had been clearly too short. Worse, the atom turned out to be much more complicated than had been thought in the early part of our century, and straightforward annihilation of a proton with an electron simply·does not work. Finally, in 1939, a much better solution was found, largely by Hans Bethe in America and Carl von

Weizsäcker in Germany. Instead of the annihilation of matter, we have the building-up of heavier elements from lighter ones; and the key to the whole situation is hydrogen.

Hydrogen is the lightest of all the elements, and is also the most plentiful, so that normal stars contain a great deal of it. Consider the Sun for a moment. Near its core, the temperature is of the order of 14,000,000 degrees, and the pressure is colossal. Strange things are happening; the nuclei of hydrogen atoms are running together to make up the nuclei of helium atoms – helium being the second lightest of the elements. It takes four hydrogen nuclei to make up one nucleus of helium, but in the process a little mass is lost and a little energy is set free. It is this energy which keeps the Sun shining – and, as we said earlier, the mass-loss amounts to 4,000,000 tons every second.

Of course, this is a very over-simplified way of putting things, and the actual helium-building process is decidedly complex, but the general principle is clear enough, so that for a Main Sequence star of solar type, hydrogen is the main 'fuel'. Now, at last, the time-scale is right. The Sun is about half-way through its stable career, and it has plenty of reserves yet.

These revelations caused a complete change of outlook with respect to the H-R Diagram, which is not an evolutionary chart as was originally thought – at least inasmuch as a star does not slip down the Main Sequence from top left to bottom right. Our Sun, for instance, is not fading away into an M-type Red Dwarf; and in the far future life on Earth will be destroyed not by cold, but by heat.

Everything in a star's career depends upon its initial mass, so let us begin with a star whose mass is about the same as that of the Sun. As it begins to condense out of the interstellar gas and dust, by gravitation, its interior becomes hot, and eventually the star starts to shine; at this stage it is unstable, so that its light fluctuates irregularly. Stars of this kind, still very young, are known as *T Tauri variables*, because the best-known member of the class is lettered T in the constellation of Taurus (the Bull).

When the central temperature has reached a certain critical value, amounting to several millions of degrees, nuclear reactions can begin. The hydrogen-into-helium process takes over; the star joins the Main Sequence, somewhere near the middle, and settles down to a long period of steady, sober existence. The Sun has been on the Main Sequence now for 4,000 million years and more, and during that time it has probably not changed much. Yet the supply of hydrogen is not inexhaustible, and at long last it will fail.

Remember, a star has to achieve a balance between the effects of energy production (tending to expand it) and the sheer weight of the outer layers (tending to make it shrink). When the hydrogen 'fuel' near the core has been used up, the star must readjust itself. Because reactions are still going on in a shell round the now helium-rich core, the outer layers will expand; but the core itself will shrink – there is nothing to balance the effects of gravitational contraction – until the inner temperature has become high enough for the helium to react in its turn. Once again energy production starts at the core, this time with helium fuel instead of hydrogen, and heavier and heavier elements are built up by a whole series of complicated processes. Meantime, the star has blown itself out into a Red Giant, with a relatively cool surface but an unbelievably hot core. It has passed into the upper right part of the H-R Diagram; and when the Sun does this there will be a period when it is emitting energy at least 100 times as strongly as it does now. Obviously, the prospects for the Earth are not very encouraging!

A star cannot stay as a Red Giant for long on the cosmical scale. There is a limit to the building-up of heavy elements from lighter ones, and so there is a limit to the production of energy. When it fails, gravitation takes over, and the star collapses into a very small, dense body of the kind known as a 'White Dwarf'. This evolutionary path is marked (a) in the diagram.

White Dwarfs are best described as bankrupt stars, with no nuclear reserves left. The best-known of them is the faint companion of Sirius, which is almost as massive as the Sun, but which has a diameter of only 26,000 miles – roughly three times that of the Earth. Packing so high a mass into so small a body means that the density must be tremendous. Fill a thimble with White Dwarf material, and the total weight would be more than a ton.

The reason why this is possible can be explained if we regard an atom as a miniature Solar System, with a central nucleus and circling electrons. The comparison is dangerous, because we now know that protons, electrons and other fundamental particles cannot be regarded as solid lumps, but it is helpful provided that it is not taken literally. An atom is largely 'waste space', and if the fundamental particles are packed together we can cram much more material into a much smaller area. This is what happens in a White Dwarf. It has reached the end of its brilliant career, and nothing lies ahead of it but eventual extinction. White Dwarfs are very common in the Galaxy, and presumably in other galaxies as well. There can be little doubt that our Sun will eventually turn into one of these cosmical glow-worms.

Next, consider a star of much lower initial mass: say less than one-tenth that of the Sun. It begins its

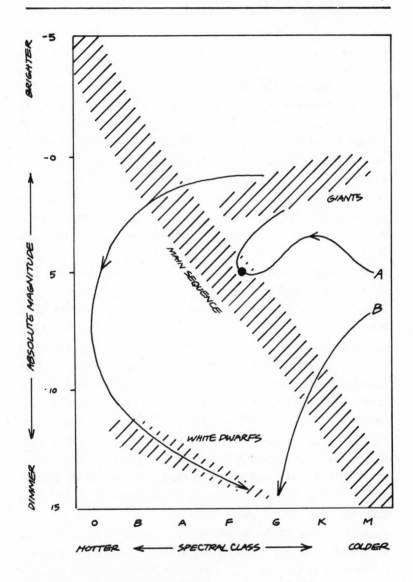

Evolutionary Tracks of Stars
(a) A star of solar mass; contraction to the Main Sequence, entry into the giant branch, collapse into a White Dwarf.
(b) A star of low mass, which evolves directly into the White Dwarf condition.

evolution in the same way, but it never becomes hot enough at its core for nuclear reactions to begin, and so it sinks straight down into the White Dwarf condition without joining the Main Sequence at all: course (b).

What, however, about a star which is much more massive than the Sun? It will evolve much more quickly; it will be much more spendthrift of its energy reserves, and instead of staying on the Main Sequence for thousands of millions of years it will remain there for only a million years or so. Rigel, for example, is probably losing mass at the rate of something like 80,000 million tons a second instead of the Sun's modest 4,000,000; and even Rigel cannot stand this depletion for long. Its life-expectancy is far less than that of the Sun. And even Rigel is not the supreme example; S Doradûs, in the Large Magellanic Cloud, is shining with at least a million Sun-power, and disaster is bound to overtake it within the next few million years at most.

It seems that if a star is over one-and-a-half times as massive as the Sun, it will not simply evolve first into a Red Giant and then into a White Dwarf. True, it will pass into the giant branch; stars such as Betelgeux are there already – and note how our ideas have changed; instead of being young, Betelgeux has been found to be decidedly senile. But when it collapses, as its fuel runs out, the reactions will become 'out of control'. With a star as massive as this, there must be a catastrophic explosion, known as a *supernova*; the star blows much of its material away into space, and the remnant shrinks down into a very small, super-dense body beside which even a White Dwarf seems tame. This time the protons and electrons run together, making up neutrons; and we are left with a *neutron star*. This time our thimble could contain something of the order of a thousand million tons!

Supernovæ are not common. During the past thousand years four have been seen in our Galaxy – in the years 1006, 1054, 1572 and 1604 – but because they are so violent, supernova outbursts can also be seen in other galaxies, and we have been able to study them. The 1054 galactic supernova, in Taurus, has left the gas-cloud we now call the Crab Nebula, together with a neutron star; long-wavelength radio emissions can be picked up from the sites of the other three galactic supernovæ.

A very massive star, then, does not go out with a whimper, but with a very pronounced bang. Suppose that the mass is greater still – say more than ten times that of the Sun? No supernova outburst will occur; when the collapse starts, nothing can stop it; and in an amazingly short time we are left with a body which is so small but so massive that not even light can escape from it. In other words, we have a Black Hole.

4 Pulsars: Mysterious Signals in the Galaxy

Look into the sky not far from the star Zeta Tauri, using a telescope of modest aperture, and you will be able to make out a faint, hazy patch. Messier saw it, and alloted it the number 1 in his list. What he did not know was that M.1, the so-called Crab Nebula, is one of the most remarkable objects in the whole sky – and one which has been of tremendous importance to astronomers. There is nothing else quite like it, so far as we know, and it is lucky that it lies at a distance of only 6,000 light-years.

The story of the Crab Nebula really began in 1054, when the old Chinese and Japanese star-gazers recorded an outburst in that part of the sky. A star blazed forth where no star had been before; it became so bright that it could be seen in broad daylight, and it remained on view for several months before it faded away. Then, naturally enough, it was more or less forgotten. Only in the age of telescopic astronomy was the Crab Nebula found, and even then it was some time before it was linked with the 'guest star' of 1054.

Lord Rosse, with his huge home-made telescope, first gave it the nickname by which we know it, and made detailed sketches of it. Later, photographs taken with powerful instruments showed the Crab Nebula to be very intricate, and to be expanding outward from what was presumably the old explosion-centre. Then, with the arrival of radio astronomy, it was found that the Crab is one of the most powerful emitters in the whole sky; moreover it also sends out X-rays, which are ultra-short, and are difficult to study because of the screening effects of the Earth's atmosphere. X-ray astronomy involves sending equipment up in rockets, which has become possible only during the past few decades.

By the 1960s it was clear that the Crab Nebula had much to offer. It produced visible light, radio waves and X-rays, so that it was energetic in all parts of the electromagnetic spectrum. There could be no doubt of its being a supernova remnant; everything fitted neatly into the pattern (though remember that the explosion did not actually occur in the year 1054; we were then seeing the results of an outburst that had occurred 6,000 years earlier, since the Crab is 6,000 light-years away). Extra evidence was provided by radio studies of other old supernovæ. In particular, the great Danish astronomer Tycho Brahe had observed a supernova in the constellation of Cassiopeia, in 1572; it also had become bright enough to be seen with the naked eye in broad daylight, and the same was true of the 1604 supernova in Ophiuchus (the Serpent-bearer) which had been studied by no less a person than Johannes Kepler, the first man to work out the way in which the planets really move round the Sun. Neither of these supernovæ had left gas-clouds comparable with the Crab Nebula, but in each case radio waves were picked up from the old sites.

During the 1960s, energetic efforts were being made to detect and catalogue the various discrete radio sources in the sky – not only the supernova remnants inside our own Galaxy, but also the radio galaxies and, of course, the quasars. Research was being followed up in many countries, notably in England, where the two leading radio astronomy observatories are at Jodrell Bank and Cambridge. It was at Cambridge that Miss Jocelyn Bell, working with the team lead by Professor Anthony Hewish, made a startling discovery. She detected a radio source which seemed to be pulsating with absolute regularity and in a very short period – like a rapid beacon. It was, rather naturally, taken to be due to Earth-based interference; but after a while the puzzled researchers realized that the signals came from something in the sky.

There was cause for bewilderment. The source was so utterly unlike anything that had been found before that there were suggestions that it might be artificial. There was a brief period when highly eminent Cambridge radio astronomers seriously entertained the idea that the 'pulsating source' or 'pulsar' might be due to signals being beamed toward us by an alien civilization far away in space. This sounds like science fiction; but not for some days was the idea finally discarded, and in the interim the Cambridge team wisely refrained from making any public announcement. (One can well imagine what the reactions of flying-saucer enthusiasts would have been – to say nothing of the daily Press, which is always eager to report any suggestion of extraterrestrial life!)

The first task was to see whether Miss Bell's pulsar was unique. It proved not to be. More were found, at Cambridge and elsewhere, and by now a hundred or so are known. Let us stress at once that despite the near-similarity in names, there is nothing in common between a pulsar and a quasar. A quasar, whatever it may be, is outside our Galaxy, and is an extremely powerful object comparable in output with a galaxy – perhaps even a hundred galaxies and more combined, if we interpret their Red Shifts as pure Doppler effects. A pulsar is contained *inside* our own Galaxy, and is relatively feeble. Probably we can detect only those which are within a few thousand light-years of us.

Once the 'alien signals' theory had been cast aside, researchers set to work to see what kind of a body a pulsar might be. Optical investigations were entirely negative. The regions from which the pulsar signals were coming gave every indication of being completely blank, so that there could not be much in the way of optical emission; and this, together with the great rapidity of the fluctuations (less than a second in most cases), made astronomers wonder whether a pulsar could be either a vibrating White Dwarf or else a neutron star. Initially, the Jodrell Bank team favoured the White Dwarf theory; the Cambridge astronomers preferred neutron stars.

The difficulty about the White Dwarf idea was the shortness of the period of the radio emissions. Although a White Dwarf is tiny compared with, say, a Main Sequence star, it is still a large object by most standards; as we have noted, the Companion of Sirius has a diameter of about 26,000 miles, while even a more extreme example, such as Kuiper's Star, is no larger than our Moon. No star of this kind could spin round at the rate of several times per second, and neither could it vibrate so quickly, so that the whole theory floundered rather helplessly. Meantime, it had been found that most of the pulsars were slowing down; the rate of increase in period was very short – a tiny fraction of a second per year – but it was detectable, because of the regularity of the signals. However, there were also variations in period which were less predictable. All this led to the idea of something which was spinning round – and it would have to be very small indeed: only a few miles in diameter at most.

It was at this juncture that pulsar-type signals were found to be coming from the Crab Nebula. The significance of this discovery could hardly be over-estimated. There was no doubt whatsoever that the Crab had to be classed as a supernova remnant, and it was also known that the explosion had been seen 900 years ago, which gave some key to the time-scale. If a pulsar were slowing down, and losing its energy, for how long could it continue to send out detectable signals? This was a matter for investigation, but certainly a pulsar would last for more than 900 years if it were the result of a supernova outburst.

The final confirmation came when the Crab pulsar was identified with a very faint, flashing object which had exactly the same period and was in exactly the right position. For the first (and, so far, the only) time a pulsar had been optically identified; and obviously it could not be a White Dwarf. It had to be a neutron star.

Rotating or vibrating? Again there were long discussions, and even now agreement between astronomers is not complete, but it looks very much as though a neutron star has a very powerful magnetic field, with the magnetic poles not necessarily coincident with the poles of rotation. Radio emissions will be sent out in the manner of the beam of a searchlight; and we will receive them only when we pass through the 'beam'. Therefore, if a neutron star is spinning round several times a second (as is quite possible for so small and dense a body) the periodical 'searchlight effect' will produce the rapidly-varying signals that we pick up. Moreover, disturbances in the neutron star itself may well cause temporary changes in the rotation period, and hence in the radio signals.

Of course, we may be wrong; but it looks as if we have at least a reliable idea of the nature of pulsars, and we are on much safer ground that we are in trying to fathom the nature of quasars, which were discovered six years earlier. The fact that not all the known supernova remnants in our Galaxy send out pulsar signals is no objection to the theory. If we do not happen to lie in the beam of the 'searchlight', nothing will be received. There is, however, a supernova remnant in the southern constellation of Vela which is associated with a pulsar; the supernova itself must have blazed forth long before there were any terrestrial astronomers to observe it, but the tell-tale gas-cloud gives us a clue. Long ago it may have looked rather as the Crab Nebula does today. Yet another gas-cloud produced by an old supernova is the so-called Cygnus Loop, in the constellation of the Swan; and here, in late 1973, S. Rappaport and his colleagues in America detected a localized source of X-rays. The evidence is mounting steadily.

A neutron star must be a weird object by any standards, and we can only speculate as to conditions there. It may have a solid crust, in which 'starquakes' occur, upsetting the regularity of the spin and causing fluctuations in the period of the radio emission which we receive. Inside a neutron star, the density must rise to values which we cannot even begin to indicate. Consider the letter 'o' on this page; can you picture a pellet with this diameter, which nonetheless weighs a thousand million tons or so?

Though a neutron star is so small, it is of tremendous mass, and so it has a very high escape velocity Nowadays, in the age of space-probes, 'escape velocity' is almost an everyday term, but it may be as well to describe it briefly just in case anyone is unfamiliar with it. Broadly, it is the velocity which an object will need to have if it is to escape from its controlling body without being given further impetus. For instance, if you could send a body upward from the Earth at a velocity of 7 miles (11 kilometres) per second, it would never come back; the Earth's gravity would not be strong enough to retain it, and the object would escape into space. From the Moon, which has a mass of only 1/81 that of the Earth, only 1½ miles (2.4 kilometres) per second are needed for escape, which is why the ascent engine in the lunar module of an Apollo spaceship can be comparatively weak. But when we come to stars, we find that the escape velocities are very high indeed.

What, then, about a neutron star? If we allow our imagination to run riot, and assume that a spacecraft could land there, it would have a difficult task to get away again, because the escape velocity will be thousands of miles per second. We are dealing with bodies which are absolutely unfamiliar in our day-to-day experience, and it is impossible to predict just what conditions there would be.

A pulsar, with its small size, rapid spin and high escape velocity, seems then to be the end-product of a very massive star, say between 1½ and 10 times as massive as our Sun. But we have already seen that with a still more massive star there will be no supernova outburst; the gravitational collapse will go on and on and on, with the dying star becoming steadily smaller. When it has passed a certain critical value in its contraction, the escape velocity from the surface will go up to a fantastic value – in fact, the velocity of light itself: 186,000 miles (or 300,000 kilometres) per second. This is the point of no return. No light can escape from the collapsing star; and since nothing can move faster than light, we have nothing more nor less than a Black Hole. So now, after this necessarily lengthy preamble, we can come to the main theme of our book.

5   The Strange Concept of the Black Hole

To say that the idea of a Black Hole is unfamiliar is a very mild way of putting matters. We have to try to visualize something which is completely outside our normal experience, and accept a situation in which we have to abandon not only the accepted laws of Nature but also what we usually call common-sense. True, such a situation is a logical follow-up of Einstein's theory of General Relativity; but to the non-specialist this is not much help either, because Relativity, too, is something beyond everyday experience.

What we hope to do, in the present chapter, is to present the Black Hole concept in a form as intelligible as is possible without delving into mathematics. We realize, only too well, that this involves some over-simplification; but it is enough if we manage to describe a picture which is not misleading. And as a start, let us bear in mind that the various models which have to be encountered cannot be taken as literally as we would like. For instance, it is very convenient to picture an atom as a sort of dwarf Solar System, with a central nucleus carrying a positive charge of electricity, and a number of circling electrons carrying negative charges. This means that each particle will be rather like a sub-microscopic billiard-ball. Modern theory stresses that one cannot regard an atom as being made up of small, solid lumps of matter; but the picture is not misleading if we are not too literal about it.

Let us, then, begin by introducing Einstein's theory of relativity – which in this context is a great deal less frightening than might be thought! It led on from the easy-to-comprehend, comforting ideas of gravity which were laid down by Sir Isaac Newton in his classic *Principia* of 1687, which has been described as the greatest mental effort ever made by one man. Newton regarded gravity as a force acting between two bodies; the Earth pulls on the Moon, the Moon pulls on the Earth, and so on. With increased distance between the two bodies, the force grows weaker according to a clear-cut mathematical relationship; thus if you double the distance, the force is reduced to one-quarter of its original value. When the theory was tested, it was found to work very well, and for most purposes it is quite adequate. For example, in calculating the orbit of a space-probe to Mars there is no need to go beyond Newton.

It was also assumed that there is 'absolute time' and 'absolute space' – so that time must pass at the same uniform rate everywhere in the universe, no matter what may be your position or the way in which you are travelling. With regard to 'absolute space', there was a practical test. We know the rate at which the Earth is moving round the sun (some 107,000 km.p.h.), but what about our velocity through space itself? As recently as the late nineteenth century it was thought that the whole of space must be filled with a mysterious medium which was called ether; light was regarded as a wave-motion, and waves have to travel in something or other (sound-waves, for instance, travel through air). If, then, we could

measure the speed of the Earth through the ether, dis-
regarding the Sun altogether, we would have something
really fundamental to guide us.

Various tests were devised, the best-known being the
Michelson-Morley Experiment of 1887. The basic idea
was that light ought to travel at a constant velocity
through the ether, and so as the Earth itself moved
through the ether it followed that the ether would seem
to stream past us. A ray of light projected in the direction
of the Earth's motion, which would thus be moving
against the stream, should take longer to cover a given
distance than ray sent out at right-angles to this direction,
i.e. across the stream. The difference between the two
speeds would provide the Earth's true velocity through
the ether. Michelson and Morley carried out the
experiment using very sensitive equipment. The results
were absolutely negative. Light seemed to travel at the
same velocity all the time, no matter at what speed the
source of light or the observer were moving.

This seemed to make no sense at all, and, inevitably,
comparisons were made with our everyday experience.
If you are walking in a corridor train, moving at 5 kilometres per hour while the train is doing 100 kilometres
per hour through a station, it seems reasonable to
assume that anyone on the platform will measure your
speed at 105 kilometres per hour – and this is the case.
All you have to do is to add the two speeds; that of the
train (100 km/h) and yourself (5 km/h), and one does not
need to be a Newton to work out that $100 + 5 = 105$.
Light travels at 300,000 kilometres per second, and the
Michelson Morley result seemed to imply that $300,000 + 300,000 = 300,000$. It was all very baffling.

It was tempting to say simply that the experiment was
faulty, and that there was some crass mistake in either the
equipment or the interpretation. Unfortunately, further

work showed that there was nothing at all the matter with the equipment, and the result had to be accepted, though explaining it was quite another problem. Common-sense had to be thrown overboard; and this is where Einsteinian relativity came in, though not, admittedly, for some years – during which period the Michelson-Morley experiment caused scientists to shake their heads in bewilderment.

Einstein's Special Theory was published in 1905. Broadly, it stated that there is no absolute standard of reference, and that everything depends upon the conditions under which the observer observes. Moreover, all interpretations are equally correct. You simply cannot take some particular object, such as the Sun, the Earth or, for that matter, Euston Station, and say: 'This is standing still; all movements must be related to it.' Everything is relative to everything else. And we are not concerned only with space and motion; we are also dealing with time. On a spacecraft rushing away from the Earth at something near the velocity of light, time will seem to slow down compared with the man who stays at home. Work up to the velocity of light itself (300,000 km/sec) and your time, relatively, will stand still. There are also very peculiar effects concerned with sheer size. The length of an object moving at such velocity will be reduced; and at 300,000 km/sec it will have been shortened to nothing at all. This is known as the Lorentz-Fitzgerald contraction, because these were the names of the two physicists who first predicted it, though admittedly for the wrong reasons.

A further effect which emerged concerns the masses of moving bodies – the faster a body travels, the more massive it becomes compared to the mass it had when standing still. In this case, the nearer that speed gets to the speed of light the more massive the body becomes until, if the speed of light could be attained, the body would have infinite mass – which is another way of saying that we can never hope to travel at the speed of light!

Quite clearly, then, we are entering a kind of Alice-through-the-Looking-Glass realm which is remarkably difficult to accept. For a long time scientists were sceptical; but various practical tests, which need not concern us at the moment, have confirmed that these weird effects do occur. Let us add that they do not become appreciable below very high velocities. There is no danger that you will contract and lose 'time' if you drive a car at, say, 150 kilometres per hour. With a spacecraft *en route* for the Moon, the contraction in length is a tiny fraction of a millimetre; *but it does happen*.

Einstein built on these ideas to develop the General Theory of Relativity, which is really a theory of gravitation – and an improvement on Newton's theory. It is often said that Einstein showed Newton to be wrong; but this is unfair – Einstein merely extended Newtonian principles, and applied them to quantities which could not possibly have been measured at the time when the *Principia* was written.

There are a few more points to be noted before we return to our Black Holes. The first is that according to relativity theory there is a very close link between space and time, which cannot really be separated. The term 'spacetime' as a frame of reference may sound meaningless, but it is not; it shows the closeness of the association. The second point is that the presence of matter introduces *curvature* to spacetime; this shows up in a number of ways. For example, a ray of light from a distant star passing close to the surface of the sun will be bent, and deflected by a small, but measurable, amount as it does so. In relativity, *gravity* is accounted for by this curvature of spacetime; the greater the curvature, the stronger the

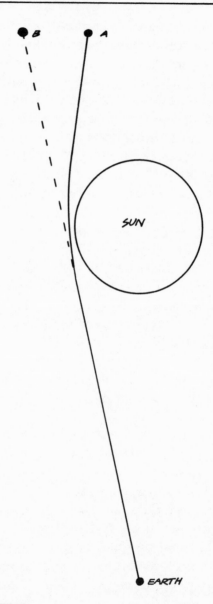

The Bending of Light

The true position of the star is (A), but due to the bending of light when passing close to the Sun, the star appears from the earth as if it is located at (B).

gravitational forces experienced.

Thirdly, there is an effect which can be traced in the spectrum of a luminous body. We have already discussed the Doppler effect, causing the spectral lines of a receding body to be shifted towards the red, or long-wave, end. Einstein predicted that there would also be a red shift due to the work which a light-wave must do when moving up in a gravitational field. To put the situation rather baldly, a light-wave escaping from a dense, massive body has to do work; this lengthens its wavelength – and we have a gravitational red shift, which has to be disentangled from any red shift caused by the Doppler principle. The gravitational red shift is most marked for a very small, very dense object; and the obvious place to look for it is in the spectrum of a White Dwarf star. When astronomers carried out the search, they found just what Einstein had predicted.

All in all, relativity theory has stood up to every test that has been devised so far; it may confound commonsense, but this cannot be helped. And when we consider anything so alien as a Black Hole, we have to reason in a relativistic world rather than a commonsensical one. Of course, there are many problems to be cleared up, and even today not all authorities are happy about details of relativity theory; but fundamentally it seems to be valid.

Now let us return to our main theme, and come to the work of the great German physicist Karl Schwarzschild. In 1916 Schwarzschild examined the case of a spherical mass of matter which was extremely small and extremely massive. It would, he said, distort spacetime so severely that nothing – not even light – could escape from it. Take the Sun, for instance. Its radius is 700,000 kilometres and as anyone can see it sends out a great deal of light. Compress the Sun, without reducing its mass, and the light will find it more and more difficult to escape. When the radius of the Sun has shrunk to a certain critical value, the light cannot get away at all. The critical radius is called the *Schwarzschild radius*,* and the boundary around the mass having this radius is called the *event horizon*, because nobody outside the boundary can have any knowledge of what is happening inside. The event horizon is an absolute barrier to any communication from within to without, and it marks the boundary of the Black Hole into which the matter has vanished. This was, indeed, the first basic concept of a Black Hole, though for many years it remained a theoretical idea only.

The term 'Black Hole' is certainly appropriate for an object such as this. Light – or matter – can fall inside it; but nothing whatsoever can escape again. To an outside observer it has disappeared with devastating permanence.

Next, let us work out the size of the Schwarzschild radius for a body the mass of the Sun. The answer is – 3 kilometres. What, then, about the Earth? If we want to turn our world into a Black Hole, we must squash it into a volume rather less than one centimetre in radius; in other words, the entire Earth must be contained in a body the size of an ordinary marble! The density of matter required defies the imagination; but, as we will see in the next chapter, there is every reason to suppose that some stars, at any rate, can end their careers by collapsing and becoming sufficiently dense to produce Black Holes.†

---

*The value of the Schwarzschild radius, Rs, is easily found for a mass M from the formula quoted in the appendix.

†It is possible to calculate the size of the Black Hole (strictly speaking a spherical, non-rotating one) which can be formed

It is only fair to point out here that, in a way, the whole idea of Black Holes was forecast by the great French mathematician, Pierre Laplace, as long ago as 1798. Thinking of light as a stream of tiny particles, and using Newton's theory of gravitation, he calculated that if a body were sufficiently dense or massive it would be invisible, because light would be unable to travel fast enough to escape from its surface. Strictly speaking, Laplace's concept was not the same as that which can be drawn from Einsteinian relativity; but it leads to much the same result, and it is interesting to find that even in Newton's theory it is possible to consider the existence of Black Hole-like objects. It is not entirely misleading to say that a Black Hole is a region of space inside which the escape velocity is greater than that of light. This is a loose way of looking at the situation, but it may perhaps make the Black Hole concept easier to visualize.

Armed with these ideas, let us now stress that a Black Hole itself is *not* a solid body. It is a region of space into which matter has fallen, and from which nothing can escape; but that matter does not fill all the volume inside the Hole. In fact, as we will see shortly, it may not occupy any volume at all. The gravitational field of the matter caused the Black Hole to form in the first place; but what happens to the matter once it is inside the event horizon has no effect at all on the size of the Black Hole. The size of the Hole depends on the *amount* of matter inside the event horizon, not the volume which it

from any mass of matter. One of the authors (I.K.N.) has calculated that his favourite politician could be contained within a Black Hole only 10^{-29} metres across, which is about one thousand million million times smaller than the nucleus of an atom. He would then, fortunately, be quite unable to communicate with the outside world!

occupies. Here again we have a situation which is hard
to credit, but which has to be accepted.

The obvious follow-up is to ask: 'If you continue to
pour matter into a Black Hole, will not the Hole eventually fill up until the surface of the mass extends outside the
Schwarzschild radius?' This sounds logical enough; but
in Black Hole studies, common-sense logic is no guide.
Let us repeat that the size of the Hole depends upon the
amount of matter inside it. If we continue to throw matter
inside, the Black Hole simply gets bigger. Its appetite,
like that of an income-tax collector, is insatiable. The
more it swallows, the bigger it gets; and in theory we
could continue 'feeding' it until it had digested all the
matter in the universe, though its basic nature would be
unaltered. It would still be a Black Hole, albeit a very
large one indeed.

Before going on to look at ways in which Black Holes
may actually form, let us think a little more about the
event horizon, which is, remember, the boundary of the
Hole. Inside the boundary, no light which is emitted can
move in an outward direction, whereas outside the
boundary some light moves outward and some falls back.
The further from the event horizon the source happens to
be, the greater the proportion of emitted light which can
avoid being sucked into the Black Hole.

It is hard to provide a good everyday analogy, but the
following may help, though again it should not be taken
too far. Imagine yourself standing by a stretch of a river
(preferably a completely smooth one) where water is
accelerating from a near-stationary state upstream
toward the lip of a waterfall downstream. Armed with a
good supply of stones, walk upstream to the stagnant
water and toss a stone in. You will see the ripples from the
splash spread out evenly upstream and downstream.
Now make your way toward the waterfall, tossing in

stones at regular intervals as you go. The nearer you get to the rim of the waterfall, the less progress upstream will the ripples make, until at last you reach a position where all the ripples are dragged downstream to the rim of the waterfall. You have, in fact, passed an 'event horizon' in the river, beyond which (from the waterfall side) no ripples can reach the world upstream. At the event horizon itself, the speed of a ripple through the water is equal to the flow of the stream, and from your point of view the ripple will stand still.

At the event horizon of a Black Hole, then, the emitted light will stand still, never moving further out into the universe and yet never falling back into the centre of the Hole. The situation for any material body, such as a spacecraft, will always be worse than this, because it cannot travel even as fast as light – let alone faster. Consequently, its chances of avoiding being sucked into the Hole are less than that of light for any given distance out from the event horizon. To extend our analogy a little further, imagine that nothing in the stream can travel through the water faster than the speed of a ripple. Clearly, even before the stream's event horizon is reached, any material body – such as a boat – will be unable to avoid the fate of being sucked on over the waterfall.

Now consider an observer on a raft, drifting downstream with the current. So far as he is concerned, he is stationary relative to the water around him; and when he encounters what we, standing on the shore, know to be the event horizon he sees what we take to be a stationary ripple pass him by, at exactly the same speed that we would see the ripples moving in the stagnant water upstream. To the man on the raft, the ripple at the event horizon is behaving perfectly normally, and is moving at the same velocity that it would do anywhere else; but to the man on the bank the ripple is stationary, and the raft passes it by on its inexorable rush to disaster. Within the context of the two respective frames of reference (raft and bank) both observers are perfectly accurate in their descriptions of what is happening. The observer on the raft notices nothing odd when he crosses the event horizon, but he can no longer communicate with anyone upstream by means of ripples. His eventual fate is sealed.

This analogy gives some idea of the differing viewpoints which would be held by a distant observer, far removed from the Black Hole, and another falling toward its centre. As the latter observer falls inward, he notices nothing out of the ordinary as he crosses the event horizon. If he were to measure the velocity of light at the boundary he would regard it as the same as usual, despite the fact that to an outsider that light would be standing still. As far as the 'falling' observer is concerned, he will drop to the centre of the Black Hole in a finite time (normally very short, depending on the size of the Black Hole). In fact he would be destroyed by tremendous gravitational tidal forces well before he reached the centre, as we will see in the next chapter; but this does not affect the general principle involved – i.e. that any particle falling into a Black Hole will reach the centre in what appears to be a finite time.

However, the outsider's view of the situation will be quite different. Earlier in this chapter we noted that according to the principles of relativity, time slows down for a rapidly-moving object. Relativity also predicts that time slows down for an observer in a strong gravitational field. To the outsider, then, the nearer the falling observer gets to the event horizon the slower time will pass for him, until at the event horizon itself time will stand still. Assume that the falling observer is illuminating himself with some very powerful torch, so that the outsider can watch his progress. The outsider will see the falling observer drawing closer and closer to the event horizon, but *never cross it*. Because the falling observer's time has slowed down to a stop so far as the outsider is concerned, as soon as he has reached the boundary of the Black Hole, the outsider will continue forever to receive light from the falling observer, whose image would be apparently frozen permanently at the event horizon.

In practice, the outsider would not be able to receive light from the event horizon in this manner, because of the phenomenon of gravitational red shift. The nearer the source of light (i.e. the falling observer) gets to the event horizon, the greater will be this red shift, and the weaker the radiation reaching the outsider. When the event horizon is reached the red shift becomes infinite and no radiation can be detected. In other words, the increasing red shift would make the falling observer fade quickly out of view before the boundary can be reached.

Nevertheless, the principle involved remains valid. So far as the falling observer is concerned, he will claim that he reaches the centre in a finite time; the outsider will say that an infinite time is needed to cross the event horizon. Which is right? The answer is: 'Both'. Remember that according to relativity there is no absolute standard of reference either for time or for space. Each observer is correct, bearing in mind his own frame of reference.

You can appreciate now our earlier comment that common-sense is no guide to what happens when we deal with these extreme conditions. It is reliable enough in everyday life – but it is quite invalid when applied to situations handled by relativity theory, and to Black Holes in particular.

6 Collapsing Stars

How can Black Holes be created in the universe? The only way known at present which could compress matter sufficiently is the phenomenon of gravitational collapse, whereby large amounts of matter may fall inwards to a common centre – due to gravitational forces acting mutually between each bit of matter involved. Take a structure such as a suspension bridge, and remove one of the supporting pillars; clearly the bridge will fall to the ground, because of the gravitational attraction of the Earth. The Earth itself is a rather rigid solid body. If, however, the material making up the Earth did not have a rigid structure preventing it from being infinitely compressed, then inexorably the effect of the terrestrial gravitational field would be to cause the Earth to collapse inwards, eventually becoming a Black Hole. Fortunately, however, the nature of matter is such that this is quite impossible.

However, if we take a spherical conglomeration of matter whose mass is a few times greater than that of the

Sun (the exact factor is a matter of debate at present), and leave it to its own devices, bereft of any means of supporting itself against gravitational attraction, then it will collapse under the effects of gravity, becoming denser and denser until an event horizon forms and a Black Hole is created. An object such as this is sometimes referred to by the appropriate term of 'collapsar'.

In Chapter 3 we looked briefly at the life-story of a star and saw the way that it evolves, how long it lives, and the nature of its ultimate fate – all of which characteristics depend very largely upon its mass. We saw that in around 5,000 to 6,000 million years' time the Sun will become a Red Giant, producing energy much more rapidly than it now does, so that it will run out of reserves of nuclear fuel fairly quickly. This process upsets the equilibrium of a star, for until then the gravitational forces tending to make it collapse have been balanced by the production of energy inside it. Remove the source of energy, and the star can no longer resist the force of gravity. Just as our bridge falls down when we remove its support, so a bankrupt star collapses when it runs out of fuel.

In the case of a star of mass similar to that of the Sun, the collapse does not continue indefinitely. Instead, it halts when a dense White Dwarf has been formed; further contraction is prevented because the electrons inside the star provide the necessary pressure. A newly-formed White Dwarf is very hot, which is of course why it merits the term 'white', but it sends out very little light compared with its former highly-luminous Red Giant state. Ultimately, a White Dwarf will cool off, eventually becoming a dark solid body called a Black Dwarf (not to be confused in any way with a Black *Hole*). If the star originally had planets travelling around it, then apart from those which were destroyed during the

Red Giant phase (as Mercury, Venus and probably the Earth will be destroyed in the case of the Sun), they will continue to orbit the dead star for the rest of time.

A typical White Dwarf is comparable in size with the Earth, and so has a radius of about 1/100 of the present radius of the Sun, or about 1/30,000 of that of the famous Red Giant star Betelgeux. If you think about the mass of the Sun compressed into a body the size of the Earth, then its density would be about a million times greater than the present density of the Sun – in other words, about a ton per cubic centimetre. This appears to be typical of White Dwarfs, and shows how matter can be compressed to densities far in excess of anything we could normally imagine.

There are many White Dwarfs to be seen in our Galaxy, and they probably represent the most common fate of stars. However, in 1931 Chandrasekhar showed that if a White Dwarf had a mass greater than 1.4 times that of the Sun, it could not – despite its great density – resist further contraction due to gravity. This limiting mass is called the Chandrasekhar Limit. (Some recent work suggests that in practice the upper limit for a White Dwarf is 1.2 solar masses, but in this text we will keep to 1.4). Thus any star whose mass is greater than this cannot become a White Dwarf unless it gets rid of its excess mass in some way. If stellar matter is compressed to even higher densities than exist in White Dwarfs, one possibility is that a neutron star will be formed – that is to say, a body in which electrons and protons (negative and positive charges respectively) have been 'squeezed' together to form neutrons (electrically neutral particles). Tremendous densities are then possible, and it seems that a typical neutron star could have a density of some 10 million tons per cubic centimetre. Because of these incredibly high densities, neutron stars are very tiny:

only about 10 kilometres in radius. If you wanted to make the Sun into a neutron star you would have to compress it to about one hundred thousandth of its present size, or, in other words, to a thousand million millionth of its present volume. The diameter of the Sun would then be no more than 10 kilometres.

The possible existence of such objects was suggested as far back as 1932 by the Russian physicist Landau, and the first detailed models of neutron stars date from the work of Oppenheimer and Volkov in the United States in 1939. Also in the United States, in 1934, Fritz Zwicky had suggested that the remnants of supernovæ might be highly compressed bodies of this kind. However, as we saw in Chapter 4, it was not until 1968 that any astronomical objects – the pulsars – were discovered whose behaviour was best explained by assuming them to be neutron stars. So far the Crab Nebula pulsar and, with less confidence, the pulsar in Vela are the only ones which have been shown to lie within supernova remnants, but a comparison of the numbers of pulsars and supernova remnants in our part of the Galaxy hints strongly that neutron stars are indeed formed from the central remnants of supernovæ.

In the White Dwarfs, it was the electrons which provided the pressure necessary to prevent further collapse of the star; in neutron stars, it is neutrons which are responsible. Despite the tremendous densities of neutron stars, there is still a maximum mass beyond which they cannot resist further contraction due to gravity. The exact value is not known with any certainty, and the whole subject is very much in the arena of debate at present, but there is little doubt that if a neutron star were more than two or three times the mass of the Sun it would collapse – and no force known to physics could stop the star's material being compressed into a smaller and smaller volume, until passing within its Schwarschild radius and forming a Black Hole.

According to current theories, what happens in the final stages of the evolution of stars much more massive than the Sun is that the central core is rather like an embryonic White Dwarf, growing inside the star. When the mass of the core exceeds the Chandrasekhar Limit, the core promptly collapses, releasing tremendous amounts of energy in a catastrophic explosion which blows the star apart. As we saw in Chapter 3, a supernova is produced; most of the matter which was contained in the star is blown away into space – and provided that the highly-compressed central remnant of the star has a mass less than two or three times that of the Sun it should form a neutron star, which is likely to be in rapid rotation.*

The amount of energy released in a supernova explosion is awe-inspiring. The star will shine as brightly as ten thousand million suns, and the total energy given out during the outburst is greater than that released by the Sun over its entire lifetime. Much theoretical work remains to be done on the mechanism responsible for supernovæ, and most mathematical models suggest that the stars blow themselves up so effectively that in most cases nothing is left to form a neutron star! However, there are good reasons for thinking that

*All stars rotate; more massive ones tend to rotate more rapidly than less massive ones such as the Sun. Consequently the central core will be rotating, and as it collapses to smaller and smaller sizes so it must rotate faster and faster. The same sort of effect can be observed with a pendulum. Set it swinging on a long string, then slide your finger down the string so that you shorten the pendulum; it will be seen to swing to and fro more rapidly before. This phenomenon is called the conservation of angular momentum.

in practice a central core will be left, though because the explosion is not perfectly symmetrical the core, and thus the neutron star, will be hurled away from the centre of the explosion. In fact, some pulsars have been found to lie at the edges of, or even beyond, the visible supernova remnants, which adds force to this suggestion.

If the mass of the central remnant of the supernova is too great then, as we have seen, the formation of a Black Hole seems inevitable. Very massive stars are likely to follow a sequence of events rather different from those which give rise to neutron stars. If the original star has a mass greater than about 10 times that of the Sun, the central core will eventually become more than two or three times the mass of the Sun, and will collapse beyond the neutron star stage, very probably forming a Black Hole within the star. The rest of the star will fall in toward the Black Hole, and may become so violently heated that much of it will be ejected in a supernova explosion. Any of the matter which crosses the boundary of the Black Hole, of course, will never escape again.

Let us now try to summarize the situation as it is understood at present. Stars with a mass of less than 1.4 times that of the Sun will probably end up as White Dwarfs. With stars up to about four times the Sun's mass, it is possible that enough mass can be shed to prevent the central core exceeding the Chandrasekhar Limit, so that such stars also will become White Dwarfs. Above four solar masses a supernova explosion is likely, leaving behind a dense central neutron star or perhaps a Black Hole. Stars with a mass more than 10 times that of the Sun seem to be the most probable candidates for the formation of Black Holes, the creation of which may or may not be accompanied by a supernova outburst. Stars as massive as this are relatively uncommon, but within our Galaxy at the moment there are enough of them to suggest that large numbers of Black Holes are likely to exist.

Let us now try to follow the fate of our collapsar as it crosses its Schwarzschild radius, with an event horizon forming around it. In other words, let us imagine that we are falling with it as it collapses, so that our time-scale agrees with that of the collapsar. As we cross the event horizon we notice nothing particularly unusual, but the gravitational field of the collapsar grows rapidly stronger and stronger. In terms of general relativity, we would say that the curvature of spacetime is rapidly increasing. Since nothing can halt the collapse, then in a very short period (measured in millionths of a second for Black Holes only a few times the mass of the Sun) all the matter is compressed to a point in the centre of the Black Hole; gravitational forces – or the curvature of spacetime – become infinite, and the matter is literally crushed out of existence. This central state of infinite spacetime curvature, called a *singularity*, where gravitational forces and density become infinite, is inescapable for a collapsing, non-rotating mass. Once inside the event horizon, nothing whatsoever can prevent the matter from ultimately meeting this fate.

The concept of matter being 'crushed out of existence' is totally alien to everyday experience. After all, we are used to the idea that matter and energy cannot be destroyed, though they can be interchanged from one form into the other; as we have noted earlier, the Sun is converting 4,000,000 tons of matter into energy every second in order to maintain its output. But everyday experience does not, fortunately, bring us into contact with infinite spacetime curvature (i.e. with singularities). In such a situation, our physical laws no longer hold good.

'Well,' it may be argued, 'if the matter is crushed out

of existence, does not this mean that the Black Hole no longer contains any matter, and must therefore vanish?' Again the answer is 'No'. This infinite curvature is frozen in spacetime inside the Black Hole – the Black Hole cannot be destroyed.

Let us think again about an outsider's view of the collapsar. Due to the time dilatation effect mentioned in the last chapter, we could see the star collapse toward its Schwarzschild radius, but never actually pass within the event horizon. The collapse would seem to slow down and stop at the event horizon; as far as we were concerned, time would stand still at the surface of the collapsar, and the image would be frozen forever at that position. Of course, we would not actually be able to observe this, because of the infinite red-shift of the light coming from the collapsar. The gravitational field of the star (and so the curvature of spacetime round the collapsar) would also be frozen in the form it had had at the instant when the collapsar reached its Schwarz- schild radius, and would remain the same for the rest of time. Nothing which happens inside the Black Hole can have any influence on what lies outside; the event horizon prohibits the passage of any kind of information from within to without. Should we ever encounter a Black Hole, then what we would notice would be the gravitational field which the body responsible for its formation had when it reached the event horizon. It is quite possible that no matter exists within it (having been crushed at the centre), but we have no way of telling so long as we sensibly decide to remain outside – and preferably well outside!

The main hazard facing any exploring astronaut approaching the vicinity of a Black Hole is the tremen- dous gravitational tidal forces which would be encoun- tered. We are all familiar with the tides in our oceans, caused by the gravitational effects of the Moon and the Sun. For the moment let us neglect the Sun, and consider only the lunar tides. It is not quite true to say that the Moon travels round the Earth; both the Earth and the Moon travel round a point lying between their centres, known as the barycentre. This is, in fact, the centre of mass of the Earth-Moon system; since the Earth is much the more massive of the two bodies, the barycentre lies much closer to the centre of the Earth. In fact, it is to be found well inside the terrestrial globe.

The essential point is that both Earth and Moon are moving under the influence of their mutual gravitational pulls; the Earth is attracting the Moon, and the Moon is attracting the Earth. Now, the water on the side of the Earth facing the Moon is closer to the Moon than is the Earth's centre, and so is subject to a slightly stronger attracting force. Similarly, the water on the Earth's far side is subject to a slightly weaker force, because it is further away than the centre of the globe. This differing gravitational pull causes the water to build up into two tidal humps on opposite sides of the Earth.

When standing on the Earth's surface, we are attracted to the Earth by gravity. However, our heads are a little further away from the centre of the Earth than our feet, and so are subject to slightly less pull. Of course this difference – this 'tidal' force between head and feet – is far too small to be noticed, but it does exist. If we were to be placed in a much stronger gravitational field, then the tidal effect would become much more obvious.

As our hypothetical astronaut approached the Black Hole, tidal effects would begin to make themselves apparent. If he approached feet-first, then he would find himself on a cosmic 'rack' of ever-increasing severity; due to the stronger gravitational attraction on his feet

compared with his head, he would be stretched out by a
rapidly-growing tidal force. For a Black Hole a few times
the mass of the Sun, as we have been describing here, the
tidal forces would kill him long before the event horizon
could be reached. The remains of the astronaut and his
spacecraft would continue to rush in to the singularity,
where they would be subject to infinite compression and
would be annihilated – all this happening in a few
thousandths of a second. The tidal force acting on a body
crossing the event horizon of a Black Hole some ten times
the mass of the Sun would be almost a hundred million
million times greater than that which we feel on the
surface of the Earth. In the vicinity of a neutron star,
the tidal force would also be enormous, making these,
too, objects to be avoided by interstellar travellers.

An interesting point about these tidal effects, however, is that the more massive the Black Hole, the less
apparent the tidal forces will become near the event
horizon. Near the *centre* of the Black Hole they still mount
to infinite strength, it is true; but further out, they diminish. The reason for this is that the strength of a tidal
force is proportional to the inverse cube of the distance
from the centre of gravitational mass. In other words,
if you increase the distance by a factor of 10, you reduce the tidal force by a factor of 1,000. Now, the
radius of a Black Hole is directly proportional to its mass,
so that a Black Hole of 100 times the mass of the Sun will
have a radius ten times greater than one which has only
10 solar masses. The event horizon of the more massive
Black Hole thus lies 10 times further away from its
centre than the event horizon of the less massive ones,
and the tidal force will be 10/1,000 = 1/100 of that
experienced at the less massive Black Hole. In short, the
tidal forces experienced when crossing the event horizon
are inversely proportional to the square of the mass of a

Black Hole. If we had a Black Hole of 100 million times the Sun's mass, the tidal force experienced by an astronaut crossing the event horizon would be completely negligible, and he could penetrate well inside the Black Hole – observing the nature of life inside – before tidal forces built up to destructive levels.

By now you may be thinking that the arguments so far advanced in this chapter have not been altogether realistic. We have been talking only about the collapse of non-rotating, spherically symmetric bodies; but *real* stars rotate, and, furthermore, the extreme conditions inside stars which lead to the formation of Black Holes are quite likely to produce a decidedly asymmetric collapsar. Surely in a realistic situation, the formation of a Black Hole is thus highly unlikely? Work is still being carried out on these problems, but it has already been shown that if the asymmetry is not great a Black Hole will still form. Likewise, rotation is not a barrier to the formation of a Black Hole unless it is excessive. It is possible that with a rapidly-rotating collapsar, the rotation might build up sufficiently to allow the surface layers of the body to exceed its escape velocity before the Schwarzschild radius is reached. If this were to happen, some of the material might be hurled away into space, but it is still very likely that most of it will go on to form a Black Hole. The possible behaviour of a rotating Black Hole has been closely investigated theoretically; we have to deal only in terms of mass, electrical charge and spin, so that the number of possible configurations is very limited. There seems to be little doubt that the spinning, collapsing core of a massive star will form a Black Hole whose properties are quite similar to those which we have described for the simple, non-rotating, spherical case we have been discussing in this and the preceding chapter.

To sum up: as the theory stands at the moment, it seems fairly certain that the fate of a really massive star (more than 10 times the mass of the Sun) is to become a Black Hole, which will continue to exist for the rest of the lifetime of the universe. The ultimate fate of the star's matter as it continues to collapse inside the Black Hole is to be crushed 'out of existence' (within any normal meaning of that term) in the spacetime singularity at the centre. Any material subsequently falling in will suffer the same fate. It is possible that the collapsing stars we have been describing in this chapter are not the only sources of Black Holes, but this question is best deferred for the moment. Our next task is to ask ourselves: Can we detect Black Holes?

7   The Search for Black Holes

The detection of Black Holes clearly presents serious problems. We cannot see them, for no light can escape from within their event horizons. Radio astronomy cannot help, because no radio waves can get out either. In fact, no signal of any kind can reach us from a Black Hole. What about radar? Again, the nature of a Black Hole rules this out. If we could send a radar beam (or, say, a laser beam) to a Black Hole, then the beam could enter it, but could never come out again; it is impossible to bounce such a signal off a Black Hole. No matter how powerful the beam, or for how long radiation is poured into the Black Hole, nothing will come back.

We are left with only two approaches to the problem. We can try to detect the gravitational field of the Black Hole (which, remember, still exists outside the event horizon, regardless of what has happened to the material inside), and we can also do our best to detect radiation being emitted by matter which is falling in towards the event horizon.

Let us think first about detecting the gravitational effects of a Black Hole. In the Galaxy there are millions of binaries, pairs of stars which revolve around each other due to their mutual gravitational pulls. The two stars making up a binary revolve around their common centre of mass, a point in space lying between them. If the two stars are of equal mass, the centre of mass lies exactly half-way between them; if one star is more massive than the other, then the centre of mass lies nearer the more massive star. As we have already noted, the same is true of the Earth-Moon system; both bodies revolve around the barycentre, which, because the Earth is 81 times as massive as the Moon, lies closer to the centre of the Earth by this same factor of 81. In fact, the barycentre lies only 5,000 kilometres from the centre of the terrestrial globe, rather more than a thousand kilometres below the Earth's surface.

Studies of binary stars are very important in astronomy, as they provide the only reliable method of measuring stellar masses. By careful measurements of the motions of the two stars, the shape and size of the binary orbit can be found, and the relative distance of each star from the centre of mass can be deduced. From this we can calculate the *relative* masses of the two stars, and if we also know the total period of time needed for the stars to complete one orbit round the centre of mass, then the combined mass of the two stars can be calculated.

Apparent Motion of Sirius
1844–1960. Instead of moving in a straight line, Sirius is made to 'wobble' because of the gravitational pull of its White Dwarf companion.

Knowing the relative masses, it is then easy enough to work out the mass of each component of the binary.

Quite often we find that one star in a binary is bright while the other is faint – too faint, perhaps, to be seen. Nevertheless, careful observation can reveal the presence of the companion star, and even allow its mass to be determined, despite its invisibility. The two components will still travel round their common centre of mass, and this will cause an apparent wobbling of the bright star as it moves through space. Many years of observation may be needed before the effects can be detected, but the technique has been tried out on various occasions with great success.

A classic case is that of *Sirius* (Alpha Canis Majoris), which is the brightest star in the sky, and is a magnificent object, visible from almost anywhere on the Earth. Its apparent magnitude is − 1.4, which makes it about a thousand times brighter than the faintest star visible with the naked eye on a really clear night. As we have noted, it is a nearby star, only 8.6 light-years from us, and with a luminosity 26 times that of the Sun. The apparent motion of Sirius across the sky was analysed by Friedrich Bessel in 1834, and it was found that instead of travelling in a straight line the star was wobbling in a period of 50 years. (This motion of Sirius from 1844 to 1960 is shown in the diagram.) Bessel concluded that an invisible companion star must be responsible for this behaviour, but no telescopes of the time were able to show it. The faint companion was eventually discovered by Clark, in 1862, while testing a new 20-inch (50-cm) refractor. Its magnitude was 8.5,* so that its luminosity

---

*Magnitude 8.5 is not particularly faint; in theory, such a star should be visible in good binoculars. However, the close proximity of the very brilliant primary tends to overwhelm

was only 1/10,000 of that of Sirius itself. Later work showed that the companion (Sirius B) is a White Dwarf; it was, in fact, the first star of this type to be recognized as such.

This technique can be used to detect the presence of very massive planets round some of the nearest stars, but the effects are very slight. If a Black Hole were a member of a binary system, however, its presence should be much more obvious. After all, the mass of a Black Hole is likely to be at least a few times that of the Sun, and may be ten, twenty or even more times as great. If a visible star were found to 'wobble' in the way we have been describing, and the mass of the invisible companion were shown to be very much greater than that of the Sun, then there would be compelling reasons for believing the companion to be a Black Hole. Any normal star several times the mass of the Sun ought to be very bright indeed; for example, a star with ten solar masses should have nearly 10,000 times the Sun's luminosity. It would be very difficult to explain a really massive, yet *invisible*, companion in any other way.

How can a normal star and a Black Hole be in orbit round each other? This is not such an improbable situation as might be thought, because it often happens that the two stars making up a binary system have very different masses. The rate at which a star evolves through its life-cycle depends primarily upon its initial mass. Massive stars are far more luminous than light-weight ones, and so they use up their nuclear fuel much faster. In this sort of situation, the less massive com-

the light from the companion, so that a large telescope is needed to show it. The two components are at their widest separation in 1975, and will then begin to move closer together in the sky until the end of the century, so that the next few years provide the best possible opportunity for seeing Sirius B.

ponent of the pair might still be a stable Main Sequence star when the other had evolved to its final state. Sirius, again, is a case in point; Sirius A, the bright component, is a Main Sequence star of type A with perhaps a thousand million years of stable existence ahead of it, while Sirius B is a dim, dying White Dwarf. Formerly, Sirius B must have been a highly-luminous massive star which evolved comparatively quickly to reach its present sad condition.

There are numerous examples of binaries in which White Dwarfs are present, and there is also good evidence to suggest that neutron stars may also be present in binary systems. During the past few years, evidence has been building up which suggests the existence of Black Holes in a few exceptional binary systems, and in one case – to be described below – the evidence is really strong. There is not as yet, however, any absolutely *conclusive* evidence that Black Holes really exist.

Before going on to describe current candidates, there is another aspect of binary-star research which is worth mentioning in some detail. Unless they are relatively close to us, binaries in which the components are close together cannot be seen as separate stars. Through the telescope we will see one starlike object only, made up of the combined light of the two members of the pair. For example, consider two stars whose distance apart is the same as the Earth's distance from the Sun. If this pair were at a distance of ten parsecs (32.6 light-years) from the Earth, then even in absolutely perfect conditions it would require a telescope of well over one metre aperture to see them as separate points of light – and at ground level on the Earth perfect conditions never occur. There are many binaries where the stars are closer than this, and, of course, most of the really interesting binaries are much more than ten parsecs away. How, then, can we discover their binary nature?

Fortunately, the spectroscope comes to our rescue. We have already described how the Doppler effect changes the apparent wavelengths of lines in the spectrum of a moving star; if the star is approaching us the lines are of apparently shortened wavelength, while if the star is receding the wavelengths seem to be lengthened. Consider a binary, with the two stars revolving round their common centre of mass, as shown in the diagram. At situation A, the two stars are moving across our line of sight, neither approaching nor receding from us (neglecting the overall towards-or-away motion of the whole system, which can be taken into account without difficulty). The spectrum we receive from the binary is made up of the combined lines from both stars, and in position A they have their normal wavelengths. By the time that the stars have moved round to situation B, one is approaching us and the other receding, and so the two sets of lines are separated out; those from the approaching star are shifted to the short-wave or violet end of the spectrum, those from the receding star shifted to the long-wave or red end. As the motions of the two stars continue, the movements of the spectral lines will repeat themselves in a regular manner. Careful measurements of these changes can often tell the astronomer a great deal about such a 'spectroscopic binary', including the masses of the stars involved.

Suppose now that one of the two 'stars' is emitting no light. The spectrum of the binary will reveal just one set of lines, from the visible star; but because of its motion round the centre of mass, the set of lines from the star will still show changes which repeat themselves periodically. Therefore, it is still possible to detect massive invisible companions even when the 'wobbling' in the star's proper motion is not directly observable. In this

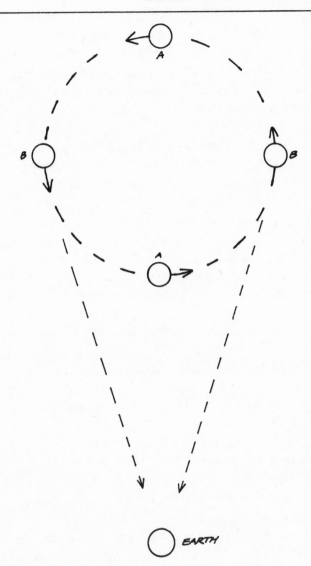

Detection of a Spectroscopic Binary
(a) The stars are moving transversely, and show no
Doppler shifts (apart from the overall shift of the system).
(b) One star is approaching, and shows a violet shift;
the other star is receding, showing a red shift, so that the
spectral lines are doubled.

situation, again, the presence of a Black Hole can be suspected if the invisible component turns out to be more than a few times more massive than the Sun.

Binaries with invisible companions are referred to as single-line binaries, because only one set of spectral lines can be seen. They are not particularly uncommon in our Galaxy, and the situation can often be explained quite simply as being due to the faintness of the secondary star; it just does not emit enough light for its spectrum to be seen. Sometimes, too, the companion may be a White Dwarf, while in other cases a neutron star would fit the facts. Investigations of these single-lines by workers such as Virginia Trimble have shown that in most cases the invisible secondary is less massive than the visible primary (and usually considerably less massive than the Sun), so that the secondaries are probably nothing more than very dim ordinary stars. There are, however, some cases in which the mass of the invisible secondary seems to be greater than that of the bright primary. If the secondary were an ordinary star, this could not be so – unless for some reason it were hidden in a cloud of obscuring material. There are, too, cases in which the mass of the secondary is several times that of the Sun. In their analysis of 1969 Trimble and Thorne were able to put forward possible explanations for each case considered which did not involve the presence of a Black Hole, and they concluded that although Black Holes might be present in some of these systems there was no compelling reason to believe that any one of them did necessarily contain a Black Hole.

Since that time, however, further cases have been found for which the case for Black Holes is stronger. At the time of writing (January 1974) the best candidate is the invisible companion of a star which is known simply by its catalogue number of *HDE 226868*. The primary is a very hot, highly-luminous supergiant star of spectral type BO. As such, it is thought to be some 30 times as massive as the Sun, and to have a surface temperature of 30,000 degrees Centigrade, against the modest 6,000 degrees for the surface of the Sun. The radius has been estimated at 23 times that of the Sun: that is to say some 18,000,000 kilometres. If HDE 226868 were placed where the Sun is now, the Earth would be bathed in radiation well over six hundred times greater than that which we actually receive, and no life as we know it would be possible.

HDE 226868 seems to be about 2 kiloparsecs or rather over 6,500 light-years away, and because of this great distance it appears as a faint star of the 9th magnitude. However, 9th-magnitude stars are visible in small telescopes, and anyone with a telescope of 2 inches (5 centimetres) aperture can see HDE 226868. It lies in the constellation of Cygnus, the Swan, not far from the 4th-magnitude star Eta Cygni, which can be seen with the naked eye.*

HDE 226868 is a single-line spectroscopic binary. Its nature has been investigated in detail by C. T. Bolton of the David Dunlap Observatory in Canada, who confirms that the rotation period around the centre of mass of the system is 5.6 days, and deduces that the mass of the invisible secondary must be about 14 times that of the Sun. There is naturally some uncertainty about the precise value, but it is fairly safe to assume that the secondary has a mass of somewhere between 10 and 20 times that of the Sun. This is much too great for either a White Dwarf or a neutron star. A Black Hole would fit the situation very well.

*For those familiar with astronomical co-ordinates: its position is R.A. 19h 56m 28s.9, declination +35°03′55″.

Further evidence comes from the new science of X-ray astronomy. Generally speaking, X-rays from distant parts of the universe cannot penetrate through to ground level because of the shielding effect of the Earth's atmosphere, and so the development of this branch of astronomy had to await the arrival of artificial satellites in orbit round the Earth, though some pioneer work was done with high-altitude rockets. The X-ray satellite, UHURU, discovered a strong source of X-rays in Cygnus, whose position was subsequently shown to coincide exactly with that of HDE 226868. This source of X-rays, known as *Cygnus X-1*, had the remarkable characteristic of rapid variation in strength; its output of X-radiation would vary even over periods as short as a tenth of a second. This short period of variation is important, as it tells us something about the size of the X-ray source. By and large it is impossible for a source of radiation to vary significantly in brightness over a period less than the time required for a ray of light to travel across the source itself. In one-tenth of a second, light can travel 30,000 kilometres, and so it would seem that Cygnus X-1 cannot be any larger than this; it may well be much smaller.

There is good spectroscopic evidence that matter is flowing from HDE 226868 to the invisible companion, and this again would be expected if a Black Hole were present. There are many examples of binaries in which one star is large and rarefied, while the other is dense and compact. If the two stars are sufficiently close together, then it is quite possible for the gravitational attraction of the second star to pull so strongly upon that part of the other component's surface immediately adjacent to it that material will be drawn out and captured by the second star. If the secondary should happen to be a Black Hole, the same effect could be expected.

Over the past year or two a rather convincing picture has been built up of how X-rays might be emitted in the situation where a star and a Black Hole are in close orbit round each other. There are several ways in which X-rays can be produced in space, and one important possibility is emission from very hot gas. In 1967 the Russian astrophysicists Shklovskii, Zel'dovich and Novikov suggested that if gas were drawn into a Black Hole, it would be heated sufficiently to emit copious quantities of X-rays. An important factor here is the sheer tiny size of the Black Hole (between 30 and 60 kilometres radius in the case of the Black Hole inside Cygnus X-1). If you try to funnel large amounts of matter torn from the surface of a star nearly twenty million kilometres across into a hole 30 kilometres across, it is easy to see that the gas will be compressed and heated to a considerable extent!

Due to the tremendous gravitational attraction of the neighbouring Black Hole, HDE 226868 is likely to be distorted into an egg-shape, with matter being dragged from the pointed end toward the Black Hole (see the diagram). What seems likely to happen, therefore, is that the material torn from the star will form a disk around the Black Hole, and that because of friction effects the gas particles on the inside of the disk will spiral into the Black Hole, with the emission of X-radiation. All the evidence seems to suggest that this is just what is happening with Cygnus X-1.

On the face of it, the case for the companion of HDE 226868 being a Black Hole is convincing, and were we dealing with a less revolutionary concept it might be taken as fairly conclusive. However, the idea of a Black Hole is fundamental in the General Theory of Relativity, and the existence of Black Holes raises so many almost bizarre implications that much more work

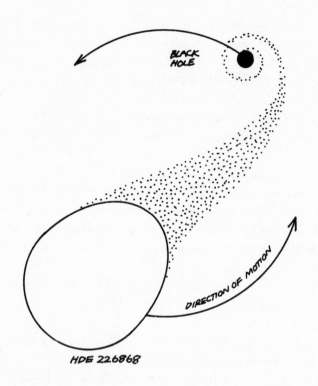

Transfer of Mass

The blue supergiant HDE 226868 and its assumed Black Hole companion are depicted here. Matter is shown streaming over from the visible star to the Black Hole. On this scale, the Black Hole should be less than a hundred millionth of a centimetre in radius.

must remain to be done before we can be entirely satisfied.

However, HDE 226868 is not the only candidate in the field, though it does seem to be the strongest. Another star which has aroused great interest in this connection is the supergiant, *Epsilon Aurigæ*, close to the brilliant Capella in the sky. It is an eclipsing binary (i.e. a binary in which each component alternately passes in front of the other, blocking out some or all of its light) with a period of 27 years. The primary is of spectral type F2 and has a mass 35 times that of the Sun, while the secondary component has been something of a mystery for many years. The primary is a true cosmical searchlight, 60,000 times as powerful as the Sun. Analysis of the spectrum of the binary usually shows only the lines due to the bright star, but during one of the rare eclipses some

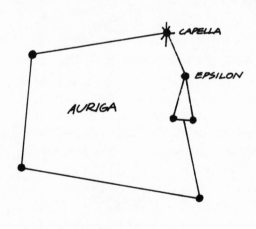

Position of Epsilon Aurigæ
Epsilon Aurigæ lies at the apex of the small triangle of stars near Capella. The three have been called the 'Hædi', or Kids.

lines caused by the secondary also show up.

The secondary is no featherweight, either – its mass is equivalent to 23 Suns! A normal star of this mass ought to be nearly half as bright as the primary. This is most certainly not the case, and clearly the secondary is a very odd object indeed. The way in which the brightness varies during each eclipse led Z. Kopal of Manchester University to suggest that the eclipse of the primary is caused by the interposition of a semi-transparent disk of material, possibly as large as 12 astronomical units in radius. If this object is regarded as a star, then it is a truly enormous one which could swallow up the orbits of all the planets in the Solar System out to and beyond Saturn. Kopal argued that both Epsilon Aurigæ and its companion must be very young, still contracting from their original gas-cloud and not yet formed into Main Sequence stars.

However, A. G. W. Cameron has pointed out that the observations do not fit this picture at all well. Instead, it seems more plausible to suggest that the visible star has evolved beyond the Main Sequence stage, while the invisible companion is a Black Hole surrounded by a disk of matter – gas and solid particles ('dust'). This dust-cloud, which seems to lie about 160 light-years from the centre of mass of the system, would be heated by radiation from the visible star, and would therefore give out infra-red radiation. Observations of Epsilon Aurigæ carried out by infra-red astronomers do indeed show that excess infra-red radiation is present, which could be explained by a large dusty cloud of particles at a temperature of a few hundred degrees Centigrade. This certainly fits in with the picture of a disk of matter round a Black Hole, but it is only right to point out that infra-red radiation is also to be expected from clouds of matter surrounding a very young embryonic star, so that the presence of infra-red radiation is not of itself proof of the Black Hole hypothesis.

Here, then, we have another plausible Black Hole candidate, though the case is less convincing than that of Cygnus X-1. Current and future observations should clarify the picture considerably. If the case for a Black Hole secondary is proved, then we will have a situation in which the effects of a Black Hole can be observed with the naked eye. Epsilon Aurigæ is normally of magnitude 3.3, and so is easily visible without optical aid. When the primary is undergoing eclipse, it fades to magnitude 4.2; that is to say the brightness is reduced to less than half its normal level. The variations are easily detectable with the naked eye, but are of course very slow. Eclipses last for over 700 days each, and occur at intervals of 27.1 years. (The next one is due in 1983.) Of course the eclipse is due to the dust-cloud; but if the Black Hole idea is correct, then the cloud can orbit as it does only because it is linked with the Black Hole. Regardless of whether its secondary is a young massive star, the collapsed remnant of an even more massive one, or something completely different, Epsilon Aurigæ is worth looking at on a clear night; its position is shown in the diagram.

Another well-known star where the case for a Black Hole has been put forward is the eclipsing binary *Beta Lyræ*. This too is a naked-eye star, lying very near the brilliant blue Vega; its magnitude ranges between 3.4 and 4.1 during its period of 12.9 days. The brightness varies continuously, so that there is no period of constant maximum interspersed with eclipses, as with Epsilon Aurigæ. This has been interpreted, quite reasonably, as showing that the two components of Beta Lyræ are so close together that they almost touch; if both components are 'stars' in the normal sense of the word, they will be distorted into egg-shapes. The possible existence of a Black Hole here was suggested because the secondary seems to be considerably fainter than the primary despite the fact that with an estimated mass of 10 Suns it is the heavier of the two components. If the secondary were a normal star it would be the more conspicuous member of the pair; actually, spectral lines from it are almost absent.

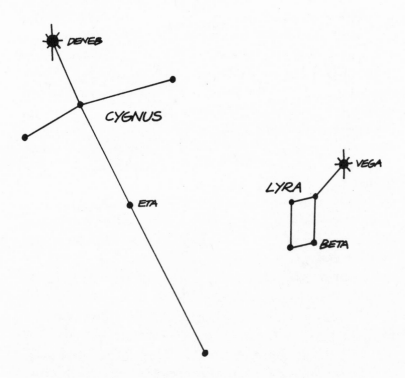

Position of Beta Lyræ

Beta Lyrae is easily recognised, near Vega. Adjoining Lyra is the prominent constellation of Cygnus, with its leader Deneb. The important star HDE 226868 lies close to Eta Cygni, which is clearly visible with the naked eye.

The Black Hole supporters argue that the secondary is a collapsar, surrounded by a cloud of matter which is spiralling in toward the event horizon. Because this material is strongly heated, it ought to radiate powerfully at short wavelengths, i.e. ultra-violet and possibly X-rays. It is also significant that ultra-violet measurements made from the highly successful satellite OAO-2 (Orbiting Astronomical Observatory 2) indicate that much of the ultra-violet from the Beta Lyræ system really is due to the secondary.

However, a much more 'normal' explanation is also available. In binary systems in which the two components are very close together, a stage can be reached where matter is transferred between the stars. For example, if one star evolves more rapidly than the other it will begin to expand as it leaves the Main Sequence, so that some of its material is dragged over to the secondary. Adding mass to the secondary in this way affects its evolution, speeding up its exit from the Main Sequence. The two stars thus lead a symbiotic existence, each influencing the other. Rapid mass-exchange between the two stars in the Beta Lyræ system could explain the faintness of the secondary; if a disk of transferred matter exists round it, the apparent faintness would not be surprising. Once again, in view of the fundamental implications of the conclusive proof of a Black Hole in a binary system, it is safer to side with the 'normal' explanation unless and until much more convincing evidence is to hand.

There are several other possible candidates under active consideration, but lest this chapter become a catalogue only one other case will be mentioned here. Another of the UHURU X-ray sources, 2U0525-06, was shown to lie close to the 5th-magnitude star Theta[2] Orionis, which is one of the stars embedded in the great

Orion nebula (M.42), a huge cloud of luminous gas easily visible to the naked eye below the three bright stars which make up the Hunter's Belt. The agreement in position is not really good, but Theta² Orionis does lie within the possible range of error in the measured position of the X-ray source. If the identification is valid, then we have a source which is basically similar to Cygnus X-1. Theta² Orionis is a very hot, luminous star, 20 times as massive as the Sun, and has an invisible companion whose mass is about 14 times that of the Sun. The orbital period is 21 days. This is longer than in the case of the Cygnus X-1 star, HDE 226868, and implies that the two bodies are further apart. The X-rays coming from this source are much weaker than those from Cygnus X-1, but due to the wider separation between the primary and the supposed Black Hole much less material would be expected to flow toward the collapsar, so that much weaker X-radiation would be just what would be expected.

The most promising line of attack in the quest for Black Holes seems to lie in investigating the binaries with massive invisible companions, particularly if these secondaries are also X-ray sources. A good theoretical picture has now been built up, showing how a collapsar in this situation could develop a surrounding disk of matter which would become exceedingly hot at the inner edge (i.e. near the event horizon), and so would emit ultra-violet or X-radiation. However, it should not automatically be assumed that binaries of this type do genuinely contain Black Holes. John Bahcall, of Princeton, New Jersey, has wisely cautioned the scientific community along the lines that we tend automatically to assume that X-ray sources in binary systems are due to matter falling in toward White Dwarfs, neutron stars or Black Holes, simply because there is no good alternative theory. He has himself suggested that some of the X-ray sources may be explained by pairs of fairly ordinary stars linked by magnetic fields. As the lines of force between the stars become twisted up, due to the rotation of the stars themselves, the surrounding matter could be strongly heated, with the resultant emission of X-radiation.

However, the quest for Black Holes is in an exciting phase. Originating as rather abstruse consequences of the General Theory of Relativity, Black Holes have begun to appear more and more plausible as the results of the collapse which marks the death-throes of a very massive star. Although they are not directly visible, we should be able to detect them by their gravitational effects on objects close to them, and also as a result of the radiation emitted from material being dragged in toward them. Observations carried out with traditional optical tele-scopes can now be correlated with the new branches of astronomy, infra-red and X-ray, in an effort to find out just where the Black Holes lie – assuming, of course, that they really exist. It is a fascinating example of the way in which science advances to consider the interplay be-tween observation and theory which is taking place in the Black Hole story. The concept itself has been dis-cussed since early in the present century, but it is only during the past few years that advances in space tech-nology have led to the development of X-ray studies – thereby pushing the relativity theorists into a frenzy to account for the results obtained. In this manner, as-tronomical observations are leading to fundamental new results which affect the most basic concepts in physics and our understanding of the universe around us.

Cygnus X-1, then, is the best candidate so far to fit the concept of a Black Hole formed from the collapsed remnants of one of the brightest and most massive stars in our Galaxy. However, in theory there is no reason why Black Holes might not arise on a much grander scale on the one hand, or on a much smaller scale on the other. These possibilities, and other aspects and implications of the Black Hole problem, are discussed in our final chapters.

8 Wider Aspects of Black Holes

So far we have been considering Black Holes formed during the final stages of evolution of stars several times the
mass of the Sun. Admittedly, some speculation is involved
in discussing such objects, but we do at least have theoretical and observational data to support us. Indeed, it
seems almost inevitable that really massive stars *must*
end up as Black Holes unless some hitherto unknown
physical process comes into play to prevent it. We also
have the behaviour of stars such as HDE 226868 to back
up our case. Yet we have taken care to stress that there
is as yet no conclusive proof of the existence of Black
Holes, though the evidence does seem to be mounting.

In recent years much theoretical work has been done,
extending the concept of the Black Hole into new fields.
Some of these fascinating speculations are described in
the following chapter. In this chapter we look at the
wider aspects and implications of the Black Hole
phenomenon.

In principle, any quantity of matter can give rise
to a Black Hole provided it can be compressed within its
Schwarzschild radius. In the case of the Sun, as we have
seen, this radius is about 3 kilometres, and the resulting
density of matter is unimaginably great – more than ten
thousand million million times greater than the Sun's
present density. Suppose, however, that a Black Hole
was formed from the collapse of a body 100 million times
more massive than the Sun. In that case, the radius of the
Black Hole would be 300 million kilometres, twice the

radius of the Earth's orbit, and the density of the collapsar when it crossed its Schwarzschild radius would be only one gram per cubic centimetre – no more than that of water. Thus, if the *mass* were great enough, a Black Hole could arise while matter still had perfectly ordinary density! In the continuing collapse, of course, the density would, in a matter of minutes, reach an infinite value.

As a further example of this, let us consider a hundred thousand million solar masses – the mass of our own Galaxy. In this case the Schwarzschild radius would be 300 thousand million kilometres, or about one-thirtieth of a light-year. Since the radius of our Galaxy is thought to be about fifty thousand light years, it would only have to be compressed by a factor of about a million to create a Black Hole. And even if this happened none of the stars in the Galaxy need be touching; they would still be quite widely separated from each other.*

It is worth considering that if our Galaxy was not rotating, it would probably have become a Black Hole a long time ago. As it is, we are saved by the motion of each star in the system balancing out the gravitational attraction towards the Galactic centre. However, the possibility that very massive Black Holes, formed from whole galaxies, may exist elsewhere in the universe cannot be ruled out, and is an idea we shall return to later in this chapter.

A further point in connection with very massive Black Holes concerns the tidal effects experienced by a body falling into them. As we saw in Chapter 6, with a Black Hole of a few solar masses these forces would

*This may help to illustrate just how widely separated are the stars in our Galaxy. At present the nearest star to the Sun is 4·2 light-years away. If the Galaxy were symmetrically compressed within its Schwarzschild radius this star would still be further from the Sun than the planet Mercury.

destroy an astronaut well before he reached the event horizon. But the tidal forces at the Schwarzschild radius diminish as the mass of the Black Hole increases. If a Black Hole comprised a whole galaxy, therefore, no discernible tidal forces would be experienced by an astronaut falling into the Black Hole at the moment he crossed the event horizon. Even so, he would only survive for a few weeks before falling inescapably into the centre and being destroyed. He would never be able to communicate his experience of 'Life inside a Black Hole' to anyone in the universe outside.

At the other end of the scale, Stephen Hawking of the University of Cambridge has speculated on the possibility of tiny Black Holes of low mass being created under certain circumstances (he was thinking particularly of the conditions which might have prevailed in the early days of the universe). Both the densities and the tidal forces at the event horizon increase with smaller mass. For example, we have already seen that the Schwarzschild radius for the Earth is about one centimetre. If compressed to that size, the Earth's density would be a hundred thousand million times greater than the Sun's density at *its* Schwarzschild radius.

The possibility of mini-Black Holes with a mass comparable to those of minor planets was taken up by Jackson and Ryan of the University of Texas, in a remarkable speculation published in *Nature*, September 1973, in which they sought to explain the 'Tungus Event' of 1908.

In June 1908 a tremendous blast occurred at Tunguska, Siberia, which flattened trees for hundreds of miles around. Eye-witnesses reported a blue streak in the sky at the time, and the sound of the blast itself was heard at great distances. The energy released may have equalled the explosion of a twenty-megaton hydrogen bomb. It has often been accounted for as the impact of a giant meteorite, but the crater and the debris which would have been expected from such a meteorite have never been found. Others have suggested it was the nucleus of a small comet which, being more diffuse than a meteorite and consisting largely of ice, might not have caused a crater. The mysterious nature of this event has not surprisingly attracted other, more extreme, explanations. It has been ascribed, for example, to the destruction of an alien spacecraft due to motor failure!

A novel suggestion made by Jackson and Ryan is that the blast was caused by the impact of a tiny Black Hole with a mass of a large asteroid. If such an entity were to exist, it would be very tiny indeed, *less than a millionth of a centimetre in radius*, but its gravitational field would be very strong. It is postulated, then, that this object approached the Earth at a speed rather greater than the Earth's escape velocity and being so massive yet so tiny passed straight through our planet and disappeared once more into space with its velocity scarcely reduced by the encounter. Their calculations indicate that the passage of such a mini-Black Hole through the atmosphere caused a shock wave with a very high temperature, giving rise to the observed streak of violet radiation and the tremendous blast at ground level. They go on to show that the object would have re-emerged in the North Atlantic Ocean somewhere around latitude 40-50 degrees north and longitude 30-40 degrees west. At the exit point there should have occurred an underwater shock, some surface disturbance, and another air shock wave. The suggestion is made that shipping records of the time should be thoroughly checked for any such reports.

This article aroused fairly heated correspondence and has been treated with scepticism. Conventional explanations, however, are not very satisfactory, and this one does at least have the merit of explaining some of the observed phenomena. Against this, it must be said that unless mini-Black Holes are very common in the Galaxy, the chances of the Earth encountering one are utterly infinitesimal and whereas the theory of Black Holes forming from collapsing stars is quite convincing the possible origin of *mini*-Black Holes is much more suspect.

Nevertheless, the idea of a mini-Black Hole is fascinating. Suppose, just for a moment, that such an object were indeed responsible for the Tungus event. What would have happened if it had been travelling slower than the Earth's escape velocity? The answer is 'Disaster!' It would have penetrated the Earth as before, but instead of retreating into space again it would have remained within our planet oscillating to and fro, digesting the Earth as it went. Eventually it would settle in the centre of the Earth and with its colossal tidal forces continue to break up surrounding matter and swallow it. The complete swallowing of the Earth would take a long time indeed due to the small size of the Hole, but the process would accelerate as its size increased. Once a Black Hole settled inside the Earth, the eventual disappearance of our planet would be inevitable!

Stephen Hawking has recently developed his theories in a way which has an interesting bearing on the concept of the mini-Black Hole. He puts forward the possibility that, in devious ways, Black Holes may in fact lose radiation from within them. If they do, then they must also lose mass. The logical conclusion to this is that Black Holes may not exist forever, as has previously been thought. Gradually losing radiation, they would eventually disappear altogether.

This suggestion is likely to be somewhat controversial, since most astronomers accept that no radiation whatsoever can escape from a Black Hole. However, let us follow Hawking's ideas to their logical conclusion. It seems the smaller the Black Hole, the greater the rate at which radiation, and hence mass, is lost. Thus, a Black Hole with the mass of our Sun would lose only a minute fraction in ten thousand million years. But if any small Black Holes had been formed early in the history of the universe with masses of only a few thousand million tons they should all have disappeared by now.

If the rate of energy loss increases with diminishing mass, it is easy to see that in the final stages a mini-Black Hole would lose radiation very rapidly. If Hawking is right, such mini-Black Holes could end up by exploding perhaps as violently as a million one-megaton hydrogen bombs. Such explosions would be tiny by astronomical standards, but fairly spectacular by comparison with what we are used to on Earth! Whether or not these ideas receive much support, we have not heard the last of the concept of the mini-Black Hole.

To return closer to the mainstream of current thinking, let us consider ways in which large Black Holes may provide the clue to some of the more mysterious energy sources in the universe. We know (subject to Hawking's theories) that we cannot receive any kind of signal from within the event horizon of a Black Hole, but we can receive energy from material falling into it. Considerable thought has therefore been given to finding ways in which such energy might in principle be extracted from Black Holes. In Chapter 3 we saw that the Sun derives its energy from the destruction of matter. About four million tons of material is converted into energy every second as a result of the nuclear reactions going on in its core. These reactions, however, are less than one per cent efficient: of every hundred tons of matter less than one ton is converted into energy. Consider now the disk of material which is thought to surround

Black Holes such as Cygnus X-1. We have already seen how it slowly spirals into the Hole, releasing energy as it goes. Calculations indicate that nearly six per cent of this material can be converted into energy. In other words, matter falling into a Black Hole may be a more efficient source of energy than nuclear fusion.

The figure of six per cent refers to a stationary Black Hole. In practice, Black Holes formed from collapsing stars should be in rotation, and spinning pretty rapidly at that. The efficiency of energy conversion would then be much greater, perhaps as high as 43 per cent! This has led people to suggest that rotating Black Holes may provide the energy source for quasars and other as yet unexplained phenomena.

The startling proposition has been made that a large Black Hole may exist at the very centre of our own Galaxy, with a mass of between ten thousand and a hundred million times that of our Sun. There does seem to be a source of energy there – strong radio and infra-red signals have been detected coming from this region.

If there is a massive Black Hole at the centre of our Galaxy, would one not be able to detect it? Surely, for example, we could see its effects on nearby stars? Unfortunately, as we saw in Chapter 2, light from the centre of the Galaxy is obscured by clouds of interstellar dust lying in the disk of our star system. Consequently, we rely upon the radio and infra-red signals to give us some idea of what lies there. These signals must have their energy source, though what it might be has puzzled astronomers for some time. Some have tried to account for the infra-red radiation by saying it is emitted by a huge cloud of dust. There still has to be some underlying energy source, however, to heat up the dust so that it gives out infra-red waves. Could a Black Hole be responsible?

Remarkable observations relating to this problem were made by Joseph Weber and his team at the University of Maryland. They found 'gravitational waves' coming from the centre of our Galaxy. Such waves can be regarded as disturbances in gravitational fields which travel at the speed of light. (Light itself is a disturbance of an electro-magnetic nature.) Suppose, for example, some agency were to annihilate the Sun. It would take just over eight minutes for knowledge of that event to reach us here on Earth. We would see the Sun continue to shine for eight minutes after the event before becoming aware it had been extinguished. Likewise, it would be eight minutes before we would notice that the Sun's gravitational field had disappeared; at which point the Earth would cease to move in a near-circular path, but would fly off at a tangent into space. In other words, the gravitational disturbance (the disappearance of the Sun) would take the same length of time as light to reach the Earth. Supernova explosions, or large masses falling into Black Holes, are clearly the sort of event likely to give rise to gravitational waves.

The detection of these waves is a very delicate task, and Weber's group was the first to claim success. His detectors consist of large aluminium cylinders. Provided the waves are powerful enough, they will cause minute vibrations in the cylinders which are then electronically amplified and recorded. The difficulty of the task, however, can hardly be overstated. Even a spectacular event such as the impact of a giant meteorite (whose volume was, say, a cubic kilometre, and whose mass was more than a thousand million million tons) would emit far too little gravitational radiation to be detected! It would, of course, produce a tremendous earth tremor and give rise to seismic waves, but that is quite a different matter. The disturbances caused by distant earth tremors or by passing traffic are far greater than gravitational waves. To help eliminate such extraneous effects, Weber has used two detectors a thousand kilometres apart. And very careful analysis of the recordings is necessary to exclude spurious results.

Weber's findings indicate that gravitational waves are reaching us from the direction of the centre of our Galaxy. So far no one else has successfully duplicated his results, and the whole idea must be treated with extreme caution. But *if* he is right, and if these waves from the Galactic centre are radiated equally in every direction, then to produce so strong a signal the Black Hole must be swallowing up thousands of stars each year! This is an alarming proposition. Such a rate of digestion could not be sustained indefinitely – at most a hundred million years, assuming that all the stars could be poured into it at a steady rate. As the Galaxy is estimated to be ten thousand million years old, either this source of gravitational waves started recently or some other explanation is called for.

One possible explanation is that gravitational radiation is only emitted in the plane of the Galaxy. This does not contradict Weber's observations, for we lie very close to this plane, but if true it would reduce the loss of mass to a much more comfortable level. Even so, for the long-term future of life on Earth, a massive Black Hole at the centre of our Galaxy would be cause for anxiety. A Black Hole can only get bigger, and it is conceivable, given a sufficiently long time-scale, that a large proportion of the Galaxy would eventually be swallowed up.

Although the suggestion of a central Black Hole can explain some observations such as the infra-red radiation (by supplying energy to heat up clouds of dust), and Weber's gravitational waves, the infra-red source can be explained in other ways, and Weber's results may in fact be wrong. It is too early, therefore, to be dogmatic about a Black Hole at the heart of our Galaxy.

It may seem a fanciful notion to suggest that entire galaxies have collapsed into Black Holes, but there is evidence for the existence of large quantities of invisible matter in clusters of galaxies. If the combined gravitational attraction of the cluster on each of its member galaxies is insufficient, then the cluster may break up. Now, there are clusters we can observe that can only be stable entities if they contain much more matter than their visible appearance suggests. The invisible matter may be in the form of non-luminous gas and dust, or galaxies too faint to be detected, but it is also possible that much of it is made up of Black Holes. There may even be more invisible than visible matter in the universe; in which case an assessment of the average density of matter in the universe is well-nigh impossible.

Black Holes have been invoked as an explanation for such enigmatic objects as the quasars described in Chapter 3. We must, of course, be wary of seeing Black Holes as the solution to all unsolved problems of energy sources in the universe. But the Black Hole hypothesis for quasars has its attractions. Let us assume the existence of a spinning Black Hole releasing up to 43 per cent of the energy equivalent of material falling into it. The enormous energy output, at this level of efficiency, could be sustained by swallowing up only one solar mass a year – a very economical process indeed compared to that required by some quasar theories.

There is another aspect of Black Holes and their central singularities which we have not so far discussed – the possibility of a 'naked' singularity. We have described how an event horizon forms when a body collapses within its Schwarzschild radius. Thereafter no signal of any kind can reach us and we cannot observe the continuing collapse of matter into a spacetime singularity. In other words, if singularities exist they are always decently cloaked by event horizons and can never be seen. There is a kind of 'cosmic censorship' which spares us embarrassment for our physical theories which become meaningless when applied to matter packed to an infinite density.

But is this cosmic censorship inevitable? It certainly seems to be, at least in the case of a *symmetrical* collapsar. It is by no means certain, however, that the collapse of a violently disturbed, or non-symmetric, object must lead to the formation of an event horizon. Indeed, if we consider a large mass rotating too fast to form a conventional Black Hole, a singularity would probably be formed in the shape of a ring instead of a point. In the plane of this ring there would be no event horizon. The singularity would be visible.

It is often said that the discovery of naked singularities would be 'disastrous' for physics; as if we had to throw the whole edifice out of the window because we could see regions of the universe in which our laws did not hold true. This is a narrow and unhistorical attitude. Physical theory has always had to respond (eventually!) to new phenomena, and it is by such response that progress has been made. If naked singularities are ever found they will provide a tremendous stimulus to theoreticians, just as the possible detection of Black Holes is doing today.

A spacetime singularity can be conceived that would function as a Black Hole in reverse. Instead of matter being crushed out of existence, it would be created. Such a singularity has been called a 'White Hole'. There is nothing in the General Theory of Relativity to say that the reverse of a Black Hole cannot occur; such entities would be singularities in spacetime out of which would pour radiation and matter that would eventually form into gas and stars.

Let us say at once that there is no evidence for the existence of White Holes, even though they have been canvassed as possible sources of the tremendous quasar energies. Nevertheless, there are the puzzling 'exploding galaxies', usually strong radio sources, which give the impression of spewing matter outwards into the universe, and it is tempting to attribute this behaviour to the existence of White Holes within them.

To take this speculation further, if – as seems to be the case – Black Holes are regions where matter disappears from our universe, then what a neat and appealing idea it would be if White Holes also existed to pour matter back into our universe again. Where would this matter come from? Two possibilities suggest themselves: either matter disappears into Black Holes and reappears elsewhere in our universe (and there are mathematical models of the nature of spacetime which would allow this to happen), or it may be that our universe coexists with other universes with which, in the normal run of things, we have no contact except via the agency of spacetime singularities. Matter disappearing through our Black Holes may emerge out of White Holes in another universe, and, *vice versa*, matter from another universe may enter ours.

We are now getting into rather deep water, but these ideas do lead us on to consider questions of cosmology, i.e. of the nature, origin, and future development of the universe itself. Consider for a moment the Big Bang theory mentioned in Chapter 2, according to which the universe began as a dense, searingly hot fireball of matter and radiation which exploded some ten thousand million years ago. Despite a certain untidiness, the Big Bang does seem the best theory of the origin of the universe and suggests that the 'creation' may in fact have been a spacetime singularity where matter poured outwards. In other words, we might regard the Big Bang as the ultimate White Hole.

Although now very much in disfavour, the Steady State theory of the universe could also call upon the White Hole concept. As we have seen, the theory implies that matter is being continually created 'out of nothing' elsewhere in the universe. It was initially pictured that matter somehow formed uniformly throughout space, but the White Hole idea would allow for matter to be created in numerous localized spots. It is an attractive idea for some cosmologists.

Since the evidence as it stands, however, favours an evolutionary universe beginning with a bang many thousands of millions of years ago, let us look ahead on this assumption and try to assess the future. There are two possibilities: either the universe continues to expand forever without limit; or the expansion eventually ceases, in which case mutual gravitational attraction will cause the universe to collapse. The galaxies will begin to move towards each other until they are all clumped together again. At this stage a new Big Bang will occur and a new expansion begin. This cyclical expansion and contraction may continue forever, and may have been going on forever in the past.

What we have been saying about Black Holes, however, raises certain problems for this 'oscillating universe' theory. If the universe does collapse and the galaxies get closer and closer together, the density of matter may increase until the galaxies merge and a spacetime singularity arises. All the matter in the universe would be infinitely compressed and crushed out of existence at the centre of a Black Hole. Far from being renewed in another Big Bang, the universe would just disappear. Naturally, we must be cautious about making such assertions. Can they be valid for the universe as a whole? If the universe is the sum of everything, can we think of it as ending in a spacetime singularity? Within what can it be singular? Can we be certain that some new and presently unknown force does not come into play just before the singular state arises to prevent collapse to infinite density? These are questions which at present we cannot answer.

There is a fascinating possibility, fascinating in a horrific sort of way, that the universe itself is a Black Hole even now! If we consider that the universe of which we are part may be a finite object in a possibly infinite spacetime, then we can imagine viewing our universe from the 'outside'. (This is not a conventional view, but the possibility has been discussed by several theorists, so let us press on). Estimates have been made as to the amount of matter in the universe, from which we can calculate its Schwarzschild radius. If the average density of the universe is about 10^{-29} grams per cubic centimetre, and the radius of the universe is about 10^{10} light-years (subject to considerable errors), we find that its Schwarzschild radius is also of the order of 10^{10} light-years! In other words, the radius of the universe is close to its Schwarzschild radius, and we may already be inside our own Black Hole.

This concept should not be taken too far; all we can say is that the numbers involved are suggestive. But there is no logical objection to the concept of universe within universe, just as there is no objection in principle to the coexistence of two or more universes with communication between them via Black and White Holes. However, having ranged from the concept of a microscopic mini-Black Hole to considering the entire universe as a Black Hole, we shall leave these speculations to be taken further by the reader.

9 Speculations

In the last chapter we looked into possible extensions of the Black Hole concept, and although observational evidence was lacking there was sound reasoning to justify our theoretical excursions. The idea of Black Holes has also given rise to wider speculation, much of it fascinating and entertaining, about the ways in which Black Holes might affect future generations of mankind. In this chapter we shall explore some of these ideas, but the reader will bear in mind that although there is a theoretical basis for the suggestions made, they are entirely hypothetical and should be treated with the proverbial grain of salt.

Navigational Hazards
It is not inconceivable that one day interstellar travel will be accomplished, and travel to distant parts of the galaxy will be regarded with no more awe than the Apollo or Skylab exploits of today. In this context it is often said that Black Holes will provide serious hazards to interstellar navigation. Obviously, it would be disastrous for a spacecraft to run into one – even if it survived entry through the Hole's event horizon it could never

escape from it, and would ultimately be destroyed.

Black Holes in close binary systems should present no problems; they would be detected by the effects on their visible companion star. Astronauts would not be able to see isolated Black Holes, however, unless they were surrounded by clouds of material falling into them. These could pose a danger. But when one considers how much empty space there is in the Galaxy, the chance of a blindfolded astronaut colliding with even an ordinary star is quite infinitesimal. If we assume that Black Holes form only from the most massive stars, and that these make up a very small proportion of the total, the chances of a collision are further reduced. We must conclude, then, that Black Holes are unlikely to constitute a threat to interstellar navigation. They would have to be deliberately sought out if spacecraft were ever to approach them.

Time dilation

A succession of close approaches to a Black Hole could allow particularly daring astronauts to achieve an almost infinite lifespan compared to their compatriots at home on Earth. We have already seen that strong gravitational fields produce a time dilation effect (see chapters 5 and 6). In the vicinity of a Black Hole time would pass more and more slowly relative to distant observers the nearer one got to the event horizon. Likewise, however, the tidal forces would build up. The astronaut would have to steer a delicate 'tightrope' course, balancing the maximum time dilation against the destructive tidal forces,* but in principle this could be done.

*The more massive the Black Hole, the smaller the tidal force at its event horizon. If these were not sufficiently obvious, the astronaut could accidentally enter the event horizon and fall inevitably to his fate at the centre.

An astronaut in this situation would not *feel* himself living any longer than his natural span. All time processes, clocks, etc., would be slowed down by the same amount in his spacecraft. But if he were to return to Earth at regular intervals between Black Hole 'trips' he would discover that much greater intervals of Earth-time, perhaps thousands of years, would have passed, and he would find a very different world to the one he left behind. (But if the present system of money-lending continued, the investment of back-pay at compound interest during these trips would reap enormous dividends!)

The question of relativity of time, raised in Chapter 5, is a fascinating one which can only be touched upon in this book. An important point is that according to the Special Theory of Relativity, time in a rapidly moving spacecraft slows down relative to a stationary, earth-bound observer's time, and in the General Theory a similar effect is produced by a strong gravitational field. Within the limits of experimental error, both concepts have been tested and verified. It really seems to happen. Most importantly, it is your own timescale that matters. If you are falling into a Black Hole, it is no comfort to recall that to an outside observer you will take an infinite time to cross the event horizon. On your timescale it will happen very quickly, and the outcome is scarcely to be contemplated.

If a future space traveller wished to extend his life-span in this way, clearly it would be a risky operation. Moreover, since he would presumably travel close to the speed of light in order to get to distant Black Holes within a reasonable time, the risks involved in a close encounter scarcely seem justified. By travelling near the speed of light he will in any case experience time dilation.

The closer he gets to the speed of light, the greater the effect.*

Should the opportunity ever arise, no doubt *somebody* would have a go! One of the greatest challenges to the individual nowadays is to sail single-handed around the world in the shortest time (with the rounding of Cape Horn providing one of the most dangerous moments). If a massive Black Hole does exist at the centre of our Galaxy, a future challenge might be to circumnavigate the Galaxy in as *long* a time (by Earth standards) as possible with the risk of crossing the Black Hole's event horizon adding more than a spice of danger to the exploit! The closer the navigator got to the event horizon, the greater the time dilation effect, but equally, the greater the danger of inadvertently entering the Black Hole.

Free energy source
We have already seen in earlier chapters that large amounts of energy, possibly in the form of gravitational waves, could be radiated away from Black Holes; for example, by matter falling in towards the event horizon. This has led to speculation that some distant day, assuming technology's ever rising demands for energy and power, the extraction of energy from Black Holes may be seriously considered. Freeman Dyson of Princeton has already pointed out the possibility that advanced civilizations might devise ways of using the entire energy

*Special Relativity shows that a spacecraft or any other material body can never travel as fast as light. Hypothetical faster-than-light particles, called 'tachyons', have been postulated although these would have quite different characteristics from ordinary matter. Recent experiments carried out in Australia suggest that these may actually have been observed. If these results are confirmed, they will raise some fascinating possibilities.

outputs of their parent stars. Vast spheres might be constructed around these stars, for example, which would collect all the emitted radiation and convert it for useful work. The 'Dyson Sphere' could supply a tremendous amount of energy at a steady rate for as long as the star continued to shine.

A method in which energy could in principle be extracted from a rotating Black Hole has been described by Dennis Sciama of Oxford University. If a rigid cubical framework were built round a rotating Black Hole, the interaction between the Hole and the cube would set the cube rotating. This would radiate gravitational waves which could then be harnessed as a source of energy. Sciama was not suggesting this would ever be practicable; but it shows another way in principle in which energy could be extracted from a Black Hole.

In a similar vein, Roger Penrose has proposed that large amounts of energy might be obtained from a Black Hole by lowering a weight into its 'ergosphere', the region of high gravitational field around the event horizon. Clearly this would require an infinitely strong rope!

Bombs
There has been a terrible inevitability about the way in which the purest and apparently least applicable of physical concepts have been used to create weapons of destruction. Relativity itself is a case in point. Einstein's relationship between mass and energy, $E = mc^2$, a result of his fine analysis of the nature of space and time, provided the basic clue to the processes involved in fission and fusion bombs (though we must not forget it also showed us how the Sun and stars shine, and how we might hope to generate large quantities of useful energy here on Earth). Likewise, a hypothetical method has already been discussed by Press and Teukolsky, of

the California Institute of Technology, by which a Black
Hole could be used as a bomb. Under certain circumstances long wavelength radio waves can be amplified in
strength by 'bouncing off' a Black Hole. The increase in
strength would be minimal, but if the Black Hole were
surrounded by mirrors capable of reflecting more than,
say, 99.8 per cent of the radiation falling upon them,
the same waves could be bounced many times off the
Hole, increasing in strength at each encounter. The
strength of the waves would be amplified exponentially
so that after a time the radiation would build up to such
a level that the mirror would be blown apart in a powerful explosion. Thankfully, however, it would be a rather
unwieldy weapon, and we need not worry unduly about
it!

10 Epilogue

We have tried to present a reasonable picture of the Black Hole concept: how such objects might be formed, how they might be detected, and what they really mean. Inevitably our account leaves much to be desired, because some of the detailed concepts are so difficult to grasp; but the Black Hole situation as it stands today may perhaps be summarized as follows.

The concept of a Black Hole arises from the General Theory of Relativity (itself the best theory of gravitation that we have at the moment) and the nature of space and time. If a certain amount of matter can be compressed within a critical radius, the Schwarzschild radius, then nothing can escape from within. A Black Hole will have been formed, the boundary of which we call the event horizon, and this Black Hole cannot be destroyed; it can only become larger, by sucking in matter. Inside the Black Hole the gravitational forces must mount up inexorably toward the centre, where the whole structure of matter as we understand it is destroyed in that region of infinite spacetime curvature which is known as a singularity. It is generally thought that we can never see these singularities, because they will always be hidden behind event horizons, but it is still open to debate whether or not detectable or 'naked' singularities could be produced under extreme conditions.

Common-sense concepts of time do not apply in the vicinity of Black Holes, and we must adjust ourselves to the strange notion that although to us outsiders a collapsing object would take an infinite time to pass through its Schwarzschild radius, the collapse to a central singularity would seem to happen quite quickly to an observer on the object itself.

We have seen, too, that gravitational collapse to form Black Holes seems to be the fate of the most massive stars in our Galaxy at the end of their life-cycles. The best chance of finding Black Holes is to look for their gravitational effects upon visible objects close to them, and the ideal situation is to look at binary stars with massive invisible companions. Several possible candidates have now been located, and are being further investigated. The present leader of the field is the invisible companion of the blue supergiant star HDE 226868, a body with about 15 times the mass of the Sun and which coincides in position with the X-ray source Cygnus X-1. The only adequate explanation of such a massive invisible body is that it is a Black Hole, and the X-ray observations fit in with this picture, since the X-radiation is thought to come from very hot gas spiralling into the Black Hole. However, the identification is not yet conclusive; alternative explanations may still be forthcoming, and a wise man will not jump too suddenly to conclusions upon a topic so startling as the discovery of a Black Hole. Nevertheless, the evidence is persuasive; and indeed, should it ever be shown that Black Holes do not occur at all, it will need some very powerful new concepts to explain their absence from the universe.

Collapsed massive stars seem to be the most promising

candidates for detection, though various researchers have suggested that Black Holes may exist in several different forms – from the very small to the very large. However, the story of Black Holes is only just beginning, and we must not regard them as being capable of providing answers to all the unsolved problems in astronomy and physics. The whole concept is truly fascinating, and has captured a wide public interest, no doubt because it is so alien to our experience in everyday life; but much remains to be learned before we can draw any really hard and fast conclusions. We will hear a great deal more about Black Holes in the years to come!

Appendix

For those who are interested in the arithmetic, the Schwarzschild radius R_s for a given mass M can be calculated from the following simple expression:

$$R_s = \frac{2GM}{c^2}$$

where G is the gravitational constant (which determines the strength of gravitational forces), and .c is the velocity of light.

Using S.I. units, where mass is measured in kilograms, length in metres, and time in seconds, $G = 6.7 \times 10^{-11}$ and $c = 3 \times 10^8$ metres per second.

For the sun, $M = 2 \times 10^{30}$ kilograms, and so

$$R_s = \frac{2 \times 6.7 \times 10^{-11} \times 2 \times 10^{30}}{(3 \times 10^8)^2} \text{ metres} = 3 \text{ kilo-}$$

metres, approximately.

This, then, is the size of the Black Hole that would be obtained if the Sun were somehow compressed sufficiently.

For comparison, the mass of the earth is 6×10^{24} kilograms.

index